Introduction to Aeroelasticity

Introduction to Aeroelasticity introduces the classical topics of aeroelasticity, beginning with elastic structural modeling and the way that wing and tail structures can diverge and deform due to aerodynamic, inertial, and control-surface deflections.

Flutter is then described, first with an airfoil example with increasing complexity, and then for flutter predictions of finite wings with and without control surfaces. Single-degree-of-freedom flutter is also discussed, which includes cable "galloping," Aeolian flutter, as well as plunging and pitching stall flutter. In all cases, the equation derivations proceed in a step-by-step fashion, complemented by several numerical (and historical) examples.

This book is intended to be a quantitative survey course for senior undergraduate or graduate students, as well as any engineer wishing a thorough introduction to the topic.

Instructors will be able to utilize a solutions manual and figure slides for their course based upon this book.

James DeLaurier has worked as an aerospace engineer, consultant, and professor. He received his PhD from Stanford University and has worked at McDonnell Aircraft, the NASA Ames Research Center, the Sheldahl Company, and Battelle Memorial Institute. Until his retirement in 2006, Dr. DeLaurier was a professor at the University of Toronto Institute for Aerospace Studies (UTIAS). His research topics have included aircraft design, lighter-than-air flight vehicles (airships and aerostats), flapping-wing aircraft, and remotely-piloted microwave-powered aircraft.

Introduction to Aeroelasticity

James DeLaurier

CRC Press
Taylor & Francis Group
Boca Raton London New York

CRC Press is an imprint of the
Taylor & Francis Group, an **informa** business

Designed cover image: April DeLaurier

First edition published 2025
by CRC Press
2385 NW Executive Center Drive, Suite 320, Boca Raton FL 33431

and by CRC Press
4 Park Square, Milton Park, Abingdon, Oxon, OX14 4RN

CRC Press is an imprint of Taylor & Francis Group, LLC

ISBN: 978-1-032-70906-2 (hbk)
ISBN: 978-1-032-70910-9 (pbk)
ISBN: 978-1-032-70909-3 (ebk)

DOI: 10.1201/9781032709093

Typeset in Times
by Apex CoVantage, LLC

Access the Support Material: www.routledge.com/9781032709062

To Dr. Ron Richardson, my dentist, who identified that I suffered from sleep apnea (when I dozed off in the dentist chair) and sent me to a sleep clinic. Since then I have enjoyed a healthier and more productive life (such as writing this book).

Contents

Preface xi

Chapter 1 Introduction 1

1.1 Collar's Triangle 1
1.2 Types of Aircraft Structures 1
1.3 Elastic Axes 2
1.4 Historical Note 4

Chapter 2 Structural Aspects: Stiffness and Flexibility 5

2.1 A Review of Fundamental Concepts and Definitions 5
2.1.1 Influence Coefficients 5
2.1.2 Strain Energy 6
2.1.3 Superposition 6
2.1.4 Maxwell's Reciprocal Theorem 7
2.1.5 Generalized Strain Energy 8
2.2 Influence Functions 12
2.3 Calculation of Influence Coefficients and Functions 13
2.3.1 Castigliano's First Theorem 13
2.3.2 Beam Equation 14
2.3.3 Torsion Member Equation 17
2.3.4 Swept Beam Equations 21
2.3.5 Differential Equation Method for Finding Influence Functions 27

Chapter 3 Divergence 28

3.1 Representative Section 28
3.1.1 Approximate Formula for a Finite Wing 29
3.2 Divergence of Large-Aspect-Ratio Unswept Wings 30
3.2.1 Integral Method, Discrete-Element Solution 33
3.2.2 Rayleigh-Ritz Method (or Assumed-Mode Method) 36
3.3 Static Divergence of Sweep Wings 38
3.3.1 Semi-Rigid Model 38
3.3.2 Determination of Equivalent-Spring Parameters F_Γ and F_θ 43
3.3.3 Discrete-Element Model 44

Chapter 4 Control Reversal 48

4.1 Introduction 48
4.2 Two-Dimensional Case 49

4.2.1 Subsonic ... 52
4.2.2 Supersonic ... 53
4.3 Cantilevered Finite Wing ... 53
4.3.1 Numerical Example ... 55
4.4 Dynamic Control Reversal ... 57
4.4.1 Steady-State Roll Rate ... 58
4.4.2 Initial Roll Acceleration ... 62
4.4.3 Numerical Example ... 65

Chapter 5 Structural Dynamics ... 67

5.1 Introduction ... 67
5.2 Torsional Vibration ... 67
5.2.1 Solution for a Uniform Cantilever Rod ... 68
5.2.2 Solution for a Uniform Free-Free Rod ... 69
5.3 Bending Vibration ... 71
5.3.1 End Conditions for Beams ... 72
5.3.2 Solution for a Uniform Cantilever Beam ... 73
5.3.3 Solution for a Uniform Free-Free Beam ... 75
5.3.4 Separate Symmetrical and Anti-Symmetrical Solutions ... 78
5.3.5 Orthogonality Relations between Modes ... 79
5.3.6 The "Work-Done" Interpretation ... 82
5.4 Approximate Methods for Modal Solutions of Bending and Torsion ... 83
5.4.1 Assumed Mode (or Rayleigh-Ritz) Method ... 83
5.4.2 Assumed-Mode Method for Combined Bending and Twisting ... 87
5.4.3 Lumped-Parameter Models ... 92

Chapter 6 Classical Flutter ... 112

6.1 Introduction ... 112
6.2 Representative Section ... 113
6.2.1 Inertial Terms ... 113
6.2.2 Elastic Terms ... 114
6.2.3 Structural Damping Terms ... 114
6.2.4 Aerodynamic Forcing Functions ... 117
6.2.5 Complete Flutter Equations ... 120
6.3 Representative Section with No Control Surface ... 121
6.3.1 Aside ... 126
6.3.2 End Aside ... 126
6.3.3 Graphical Summary ... 128
6.3.4 Numerical Examples ... 130
6.3.5 Further Examples from Aeroelasticity Literature ... 136
6.3.6 An Energy Interpretation of Flutter ... 139

6.4 Finite Wing 140
6.4.1 Uncoupled Assumed Modes 140
6.4.2 Discrete-Element Numerical Example 144
6.4.3 Coupled Assumed Modes 148
6.4.4 Quasi-Normal Modes 154

Chapter 7 Single-Degree-of-Freedom Flutter 157

7.1 "Galloping" Flutter 157
7.2 Aeolian Flutter 159
7.3 Plunging Stall Flutter 160
7.3.1 Numerical Example 161
7.4 Pitching Stall Flutter 163
7.4.1 Numerical Example 164

Appendix A Lumped-Parameter Coupled Bending and Twisting 165
A.1 Unswept Wing 165
Numerical Example 1 167
Numerical Example 2 168
A.2 Swept Wing 170
Numerical Example 3 172

Appendix B Flutter of the Representative Section with Control Surface 174
Numerical Example 177
Case 1: Variation of Control-Spring Stiffness 178
Case 2: Variation of Control-Surface Mass 178
Case 3: Location of Control-Surface's Mass 179

Index 181

Preface

The purpose of this book is to present an introductory course in aeroelasticity to aerospace students in graduate school or their fourth year of undergraduate studies. It is recognized that this may be their only formal exposure to the subject, so the structure of the course is broad-based: a quantitative survey course to give the scope and flavor of the subject. Also, this work is presented more as a series of enhanced lecture notes than as a formal reference text. Therefore, these may be readily used as a teaching source.

This book evolved from a course first taught by Prof. R. C. Tennyson of the University of Toronto Institute for Aerospace Studies and Mr. John Glaser, who was Chief Aeroelastician at De Havilland Canada. The author inherited the course in 1975 and subsequently developed it into its present form. Also, as computing capability dramatically increased over the years, it became convenient to perform numerical solutions with MATLAB programs. The author is grateful for help with this from his graduate students, especially Mr. Ben Cho. After my retirement in 2006, the course was passed on to Prof. Peter Grant of the Institute for Aerospace Studies, and he carefully went through the notes, correcting bloopers and numerical errors. This book has greatly benefitted from his scrupulous work. However, please note that any remaining mistakes are my responsibility.

Finally, I very much appreciate Prof. Holt Ashley's permission to use several examples from his two classic books on aeroelasticity. Prof. Ashley was my thesis advisor at Stanford many years ago, and the clarity of his teaching and mentoring has been an inspiration throughout my career.

James DeLaurier,
Toronto, Canada, 2024

1 Introduction

1.1 COLLAR'S TRIANGLE

Three fundamental disciplines in aerospace engineering are aerodynamics, rigid-body dynamics, and structural elasticity. These may be represented by the three circles in the diagram shown as follows (first presented by Arthur R. Collar and known as "Collar's triangle"). The interaction between and among them represent inter-disciplinary fields, including those of aeroelasticity.

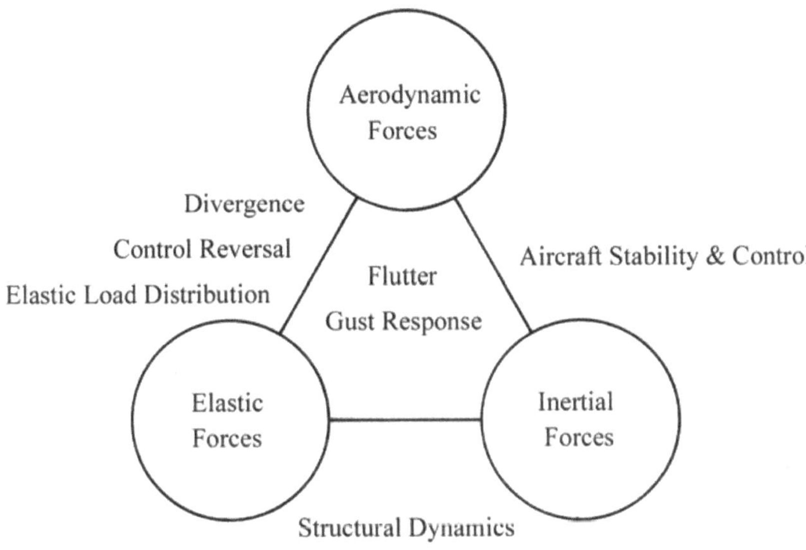

In particular, the interaction between aerodynamic and elastic forces gives rise to three topics in "static aeroelasticity": wing divergence, control reversal, and the modification of structural loads caused by deformation. Further, topics in "dynamic aeroelasticity," such as flutter and gust response, involve an interaction of all three fundamental disciplines. Note that if control theory were also included as a circled topic, the involvement of all four would be the field of "aeroservoelasticity."

1.2 TYPES OF AIRCRAFT STRUCTURES

The aircraft considered in this course will be "conventional," in the sense that they are comprised of sufficiently slender elements so that these components may be treated as beams.

DOI: 10.1201/9781032709093-1

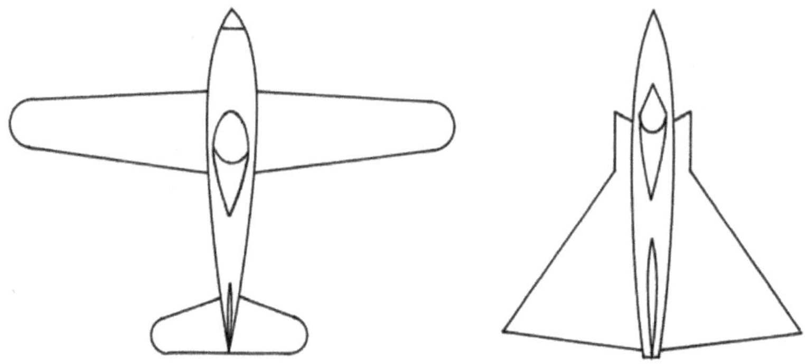

Conventional

- High Aspect Ratio
- Slender fuselage
- All components modelled as beams.

Unconventional

- Wing planform may have span- and chordwise bending of equal magnitudes. Finite element program is required.

"Unconventional" aircraft do not allow the beam assumption to be made, but the principles described in this course may be extended to treat these cases.

1.3 ELASTIC AXES

For every elastic beam cross section, there exists an "elastic center" such that the cross section translates, but does not rotate, under an applied load.

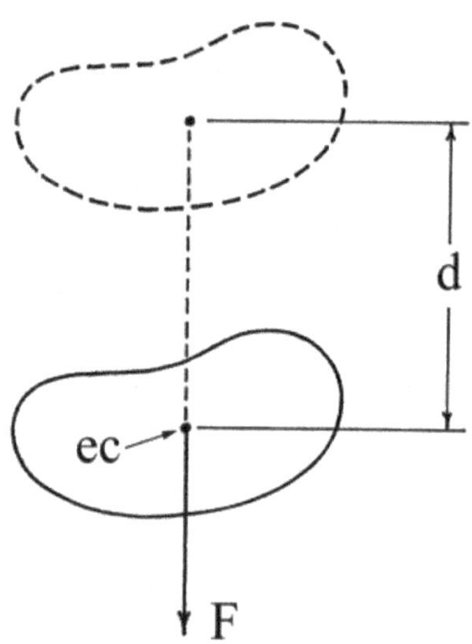

Further examples of elastic centers are shown as follows. Note that, in general, the elastic center does not coincide with the geometrical center.

The locus of elastic centers on a beam structure is called the "elastic axis."

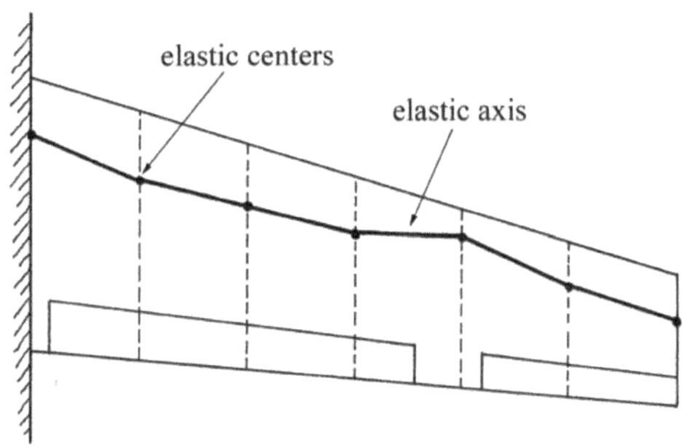

Structurally, aircraft in this course will be considered to be composed of elastic axes:

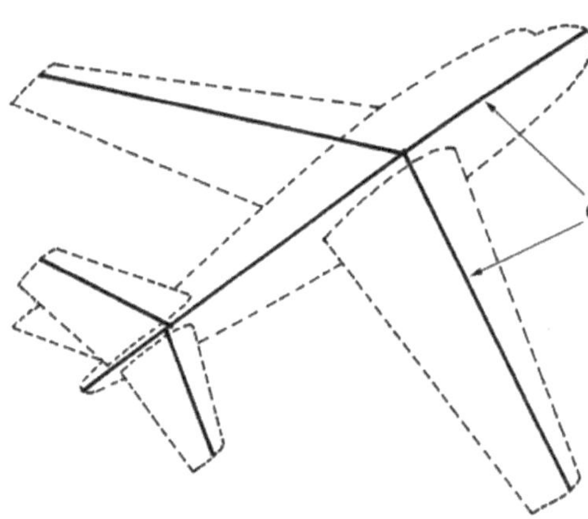

The wing is assumed to have no chordwise bending.

1.4 HISTORICAL NOTE

Aeroelastic considerations have existed from the dawn of airplane flight. The preceding picture shows the launch of the Langley Aerodrome in 1903. This was a piloted engine-powered aircraft that was the end result of a long development program involving successful scale models and aerodynamic testing. However, the full-scale aircraft crashed because of excessive aeroelastic deformation that caused the aircraft's trim state to be unstable. The strong negative twist of the front wing can be clearly seen. Shortly afterward, the Wright brothers achieved successful flight with a biplane design that had sufficient structural stiffness and strength.

2 Structural Aspects
Stiffness and Flexibility

2.1 A REVIEW OF FUNDAMENTAL CONCEPTS AND DEFINITIONS

The following concepts are important for defining and quantifying the characteristics of a structure:

Strength: the ability to carry a load.
Stiffness: the ability to resist deformation.
Linearity: that, for elastic structures, all forces and displacements are linearly related.

2.1.1 Influence Coefficients

Consider the cantilever beam example shown. The deformation under load is given by

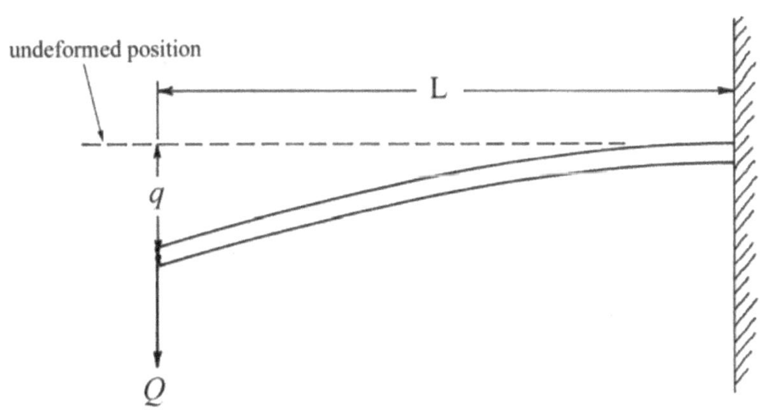

$$q = \left(\frac{L^3}{3EI}\right) Q \equiv C\,Q, Q = K\,q$$

where
$C \equiv$ flexibility influence coefficient (displacement/unit load)
$K \equiv$ stiffness influence coefficient (loading for unit displacement)

DOI: 10.1201/9781032709093-2

2.1.2 Strain Energy

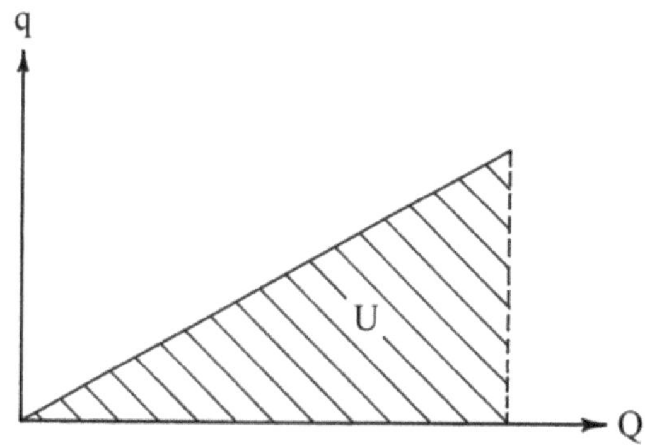

With gradual deformation under applied load, the work done equals the increase in internal strain energy for linear structures. For the cantilever beam example, this is given by the area under the $Q-q$ curve:

$$U=\frac{1}{2}Qq=\frac{1}{2}CQ^2=\frac{1}{2}Kq^2$$

2.1.3 Superposition

Consider the following example:

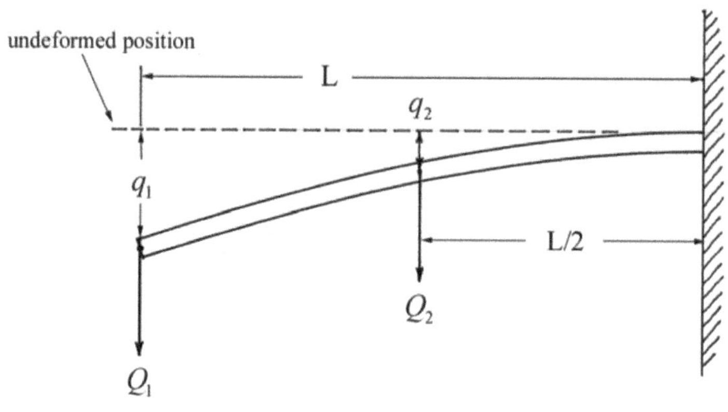

$$q_2=\left(\frac{L^3}{24EI}\right)Q_2+\left(\frac{5L^3}{48EI}\right)Q_1=C_{22}Q_2+C_{21}Q_1 \tag{2.1}$$

In general,

$$q_i=\sum_{j=1}^{n}C_{ij}Q_j \text{ or } \{q\}=[C]\{Q\} \tag{2.2}$$

where

$$\begin{bmatrix} C_{11} & C_{12} & . & . & . & C_{1n} \\ C_{21} & C_{22} & . & . & . & C_{2n} \\ . & . & . & . & . & . \\ . & . & . & . & . & . \\ . & . & . & . & . & . \\ C_{n1} & C_{n2} & . & . & . & C_{nn} \end{bmatrix}$$

The jth column of this matrix describes the deformation shape due to a unit value of Q_j, with all other $Q's$ equal to zero.

One can also obtain a stiffness-influence coefficient representation:

$$Q_i = \sum_{j=1}^{n} K_{ij}\, q_j \text{ or } \{Q\} = [K]\{q\} \tag{2.3}$$

where

$$\begin{bmatrix} K_{11} & K_{12} & . & . & . & K_{1n} \\ K_{21} & K_{22} & . & . & . & K_{2n} \\ . & . & . & . & . & . \\ . & . & . & . & . & . \\ . & . & . & . & . & . \\ K_{n1} & K_{n2} & . & . & . & K_{nn} \end{bmatrix}$$

The jth column of this matrix gives the load distribution required to produce a unit displacement at station j, with zero displacements at all other stations. Also, a comparison of Equations (2.2) and (2.3) gives:

$$[K] = [C]^{-1} \text{ and } [C] = [K]^{-1}$$

2.1.4 Maxwell's Reciprocal Theorem

For the uniform cantilever beam, one may obtain that

$$q_1 = \left(\frac{5L^3}{48\,EI} \right) Q_2 = C_{12}\, Q_2$$

A comparison with Equation (2.1) gives that

$$C_{12} = C_{21}$$

which may be generalized to the following relationships for all linear structures:

$$C_{ij} = C_{ji},\ [C] = [C]^T,\ K_{ij} = K_{ji},\ [K] = [K]^T$$

2.1.5 Generalized Strain Energy

In a similar fashion, the strain energy for a linear structure may be generalized to give

$$U = \frac{1}{2}\sum_{i=1}^{n} Q_i\, q_i = \frac{1}{2}\sum_{i=1}^{n}\sum_{j=1}^{n} K_{ij}\, q_j\, q_i = \frac{1}{2}\sum_{i=1}^{n}\sum_{j=1}^{n} C_{ij}\, Q_j\, Q_i$$

$$U = \frac{1}{2}\, Q\, \{q\} = \frac{1}{2}\, q\, [K]\{q\} = \frac{1}{2}\, Q\, [C]\{Q\}$$

Aside: Note the matrix symbol convention:

$$[\,] \equiv SquareMatrix; \{\,\} \equiv ColumnMatrix; \equiv RowMatrix$$

2.1.5.1 Example 2.1

Consider the uniform cantilevered rod with evenly spaced nodes, as shown. The terms for the flexibility influence coefficient matrix may be found as follows:

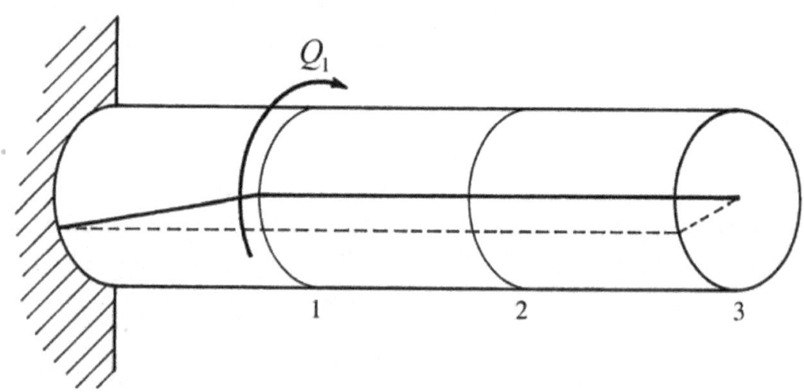

For this first case, the torque is applied at station 1, which gives the deformations at all three stations as

$$q_1 = \tilde{C}\, Q_1,\ q_2 = q_3 = q_1$$

$\tilde{C}$ is an arbitrary constant that may be adjusted to give a unit twist at station 1 in response to a unit applied torque. Therefore, the flexibility influence coefficients for this case may be expressed as

$$C_{11} = \tilde{C},\ C_{21} = \tilde{C},\ C_{31} = \tilde{C}$$

In a similar fashion, the second loading case gives:

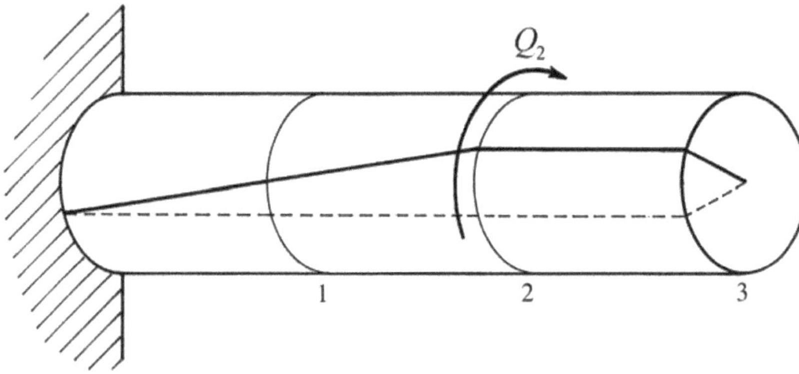

$$q_2 = 2\tilde{C}Q_2, q_1 = \tilde{C}Q_2, q_3 = 2\tilde{C}Q_2$$

$$C_{22} = 2\tilde{C}, C_{12} = \tilde{C}, C_{32} = 2\tilde{C}$$

The results for the third loading case are:

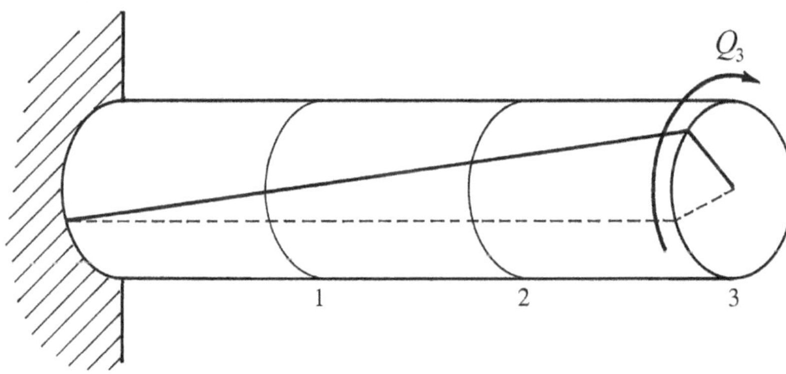

$$q_3 = 3\tilde{C}Q_3, q_2 = 2\tilde{C}Q_3, q_1 = \tilde{C}Q_3$$

$$C_{33} = 3\tilde{C}, C_{23} = 2\tilde{C}, C_{13} = \tilde{C}$$

In summary, the flexibility influence coefficient matrix is

$$[C] = \begin{bmatrix} \tilde{C} & \tilde{C} & \tilde{C} \\ \tilde{C} & 2\tilde{C} & 2\tilde{C} \\ \tilde{C} & 2\tilde{C} & 3\tilde{C} \end{bmatrix}$$

2.1.5.2 Example 2.2

The stiffness influence coefficients for the rod are now found. The first column of the matrix is obtained from the following loading sequence:

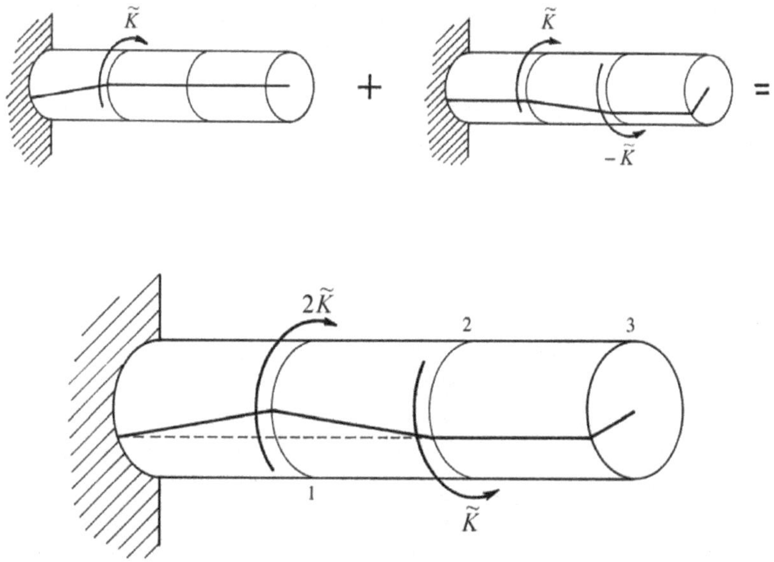

This is the loading that gives a displacement at station 1, with zero displacement at stations 2 and 3. Further, $\tilde{K}$ may be chosen to make the displacement equal to unity. Assuming this is done, one obtains

$$K_{11} = 2\tilde{K},\ K_{21} = -\tilde{K},\ K_{31} = 0$$

The next column is found from the following loading sequence:

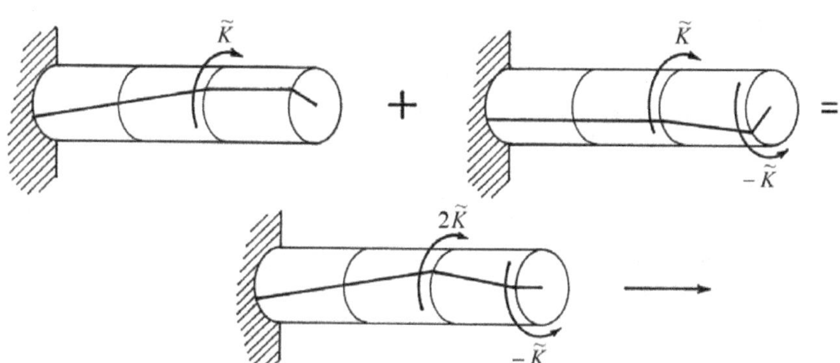

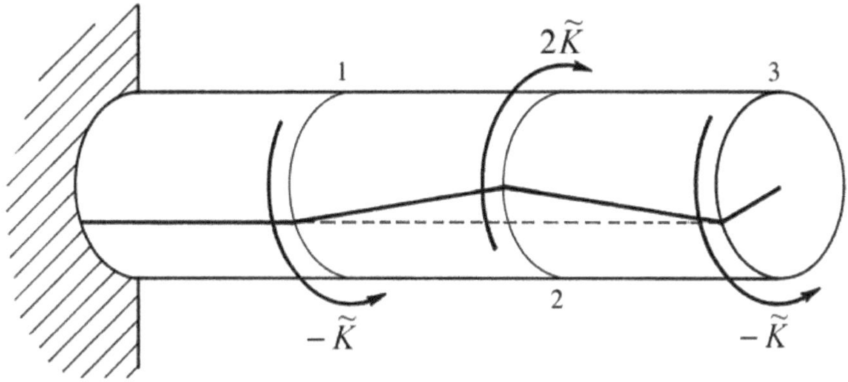

which gives

$$K_{12} = -\tilde{K}\ ,\ K_{22} = 2\tilde{K}\ ,\ K_{32} = -\tilde{K}$$

The loading for the last column is shown as follows:

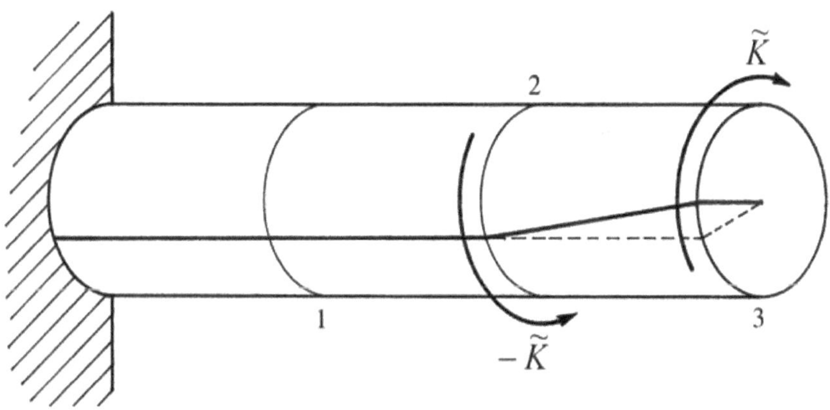

$$K_{13} = 0,\ K_{23} = -\tilde{K},\ K_{33} = \tilde{K}$$

In summary, the stiffness influence coefficient matrix is

$$[K] = \begin{bmatrix} 2\tilde{K} & -\tilde{K} & 0 \\ -\tilde{K} & 2\tilde{K} & -\tilde{K} \\ 0 & -\tilde{K} & \tilde{K} \end{bmatrix}$$

2.2 INFLUENCE FUNCTIONS

The influence function concept is an extension of the influence coefficient definition:

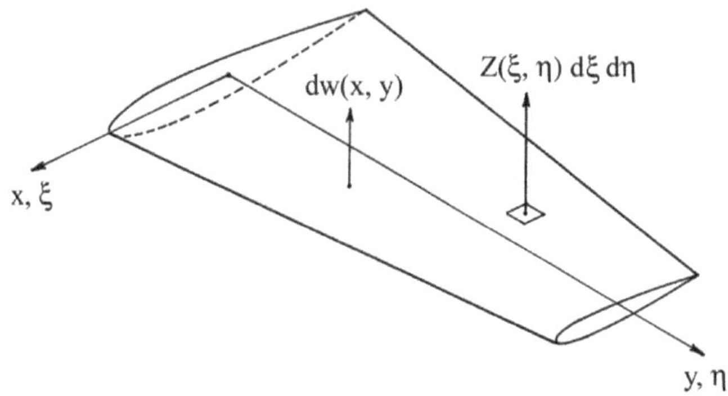

$$w(x,y)=\iint_{surface} C(x,y:\xi,\eta)Z(\xi,\eta)d\xi\, d\eta$$

where

Z is a force/area, w is a displacement, $C(x,y:\xi,\eta)=C(\xi,\eta:x,y)$

In accordance with the chordwise rigid structural assumption stated previously, this may also be defined with respect to the elastic axis:

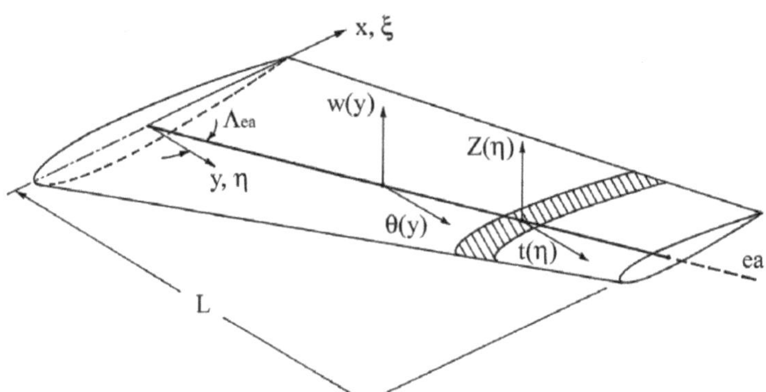

$$w(y)=\int_0^L C^{zz}(y,\eta)Z(\eta)d\eta+\int_0^L C^{z\theta}(y,\eta)t(\eta)d\eta \tag{2.4}$$

$$\theta(y)=\int_0^L C^{\theta z}(y,\eta)Z(\eta)d\eta+\int_0^L C^{\theta\theta}(y,\eta)t(\eta)d\eta \tag{2.5}$$

where

$$Z(\eta) = \int_{chord} Z(\xi,\eta)\,d\xi = \text{ load/span}$$

$$t(\eta) = \int_{chord} (\xi - \eta\tan\Lambda_{ea})Z(\xi,\eta)\,d\xi = \text{torque/span}$$

$C^{zz}(y,\eta) \equiv w$ displacement of elastic axis at y due to a unit load (normal to wing) at η.

$C^{z\theta}(y,\eta) \equiv w$ displacement of elastic axis at y due to a unit torque at η.

$C^{\theta z}(y,\eta) \equiv \theta$ displacement of elastic axis at y due to a unit load at η.

$C^{\theta\theta}(y,\eta) \equiv \theta$ displacement of elastic axis at y due to a unit torque at η.

Note that

$$C^{z\theta}(y,\eta) = C^{\theta z}(\eta, y) \neq 0 \text{ for swept wings}$$

If $\Lambda_{ea} = 0$, then $C^{z\theta} = C^{\theta z} = 0$.

2.3 CALCULATION OF INFLUENCE COEFFICIENTS AND FUNCTIONS

Energy methods are used to determine these values:

2.3.1 Castigliano's First Theorem

The deflection of any *elastic* structure in the direction of the load acting on it is equal to the partial derivative of the complementary energy U^* with respect to the load (proof of this theorem may be found in *The Analysis of Structures* by N. J. Hoff, Wiley, 1956, pp. 377–378):

$$q_i = \frac{\partial U^*}{\partial Q_i}$$

where

$$U^* = \int_{Vol} dU^* = \int_{Vol}\left(\int_0^{\sigma_1} \epsilon\, d\sigma + \int_0^{\tau_1} \gamma\, d\tau\right) d(Vol)$$

In comparison, the strain energy is given by

$$U = \int_{Vol} dU = \int_{Vol}\left(\int_0^{\epsilon_1} \sigma\, d\epsilon + \int_0^{\gamma_1} \tau\, d\gamma\right) d(Vol)$$

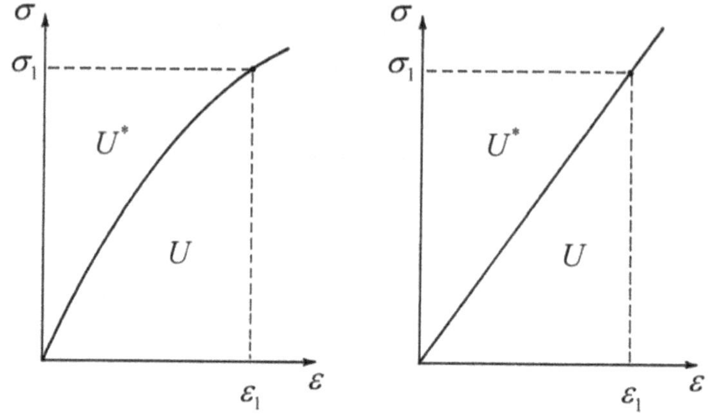

It can be seen that for *linear* elastic structures, $U^* = U$. Therefore, one obtains:

$$q_i = \frac{\partial U^*}{\partial Q_i} = \frac{\partial U}{\partial Q_i} \tag{2.6}$$

Also, from the definition of flexibility influence coefficients, as well as the expansion of a function by partial derivatives, one has

$$q_i = \sum_{j=1}^{n} C_{ij} Q_j = \sum_{j=1}^{n} \frac{\partial q_i}{\partial Q_j} Q_j$$

which gives

$$C_{ij} = \frac{\partial q_i}{\partial Q_j} = \frac{\partial^2 U}{\partial Q_j \partial Q_i} = \frac{\partial^2 U}{\partial Q_i \partial Q_j} \tag{2.7}$$

This may likewise be extended to flexibility influence functions:

$$C^{zz}(y,\eta) = \frac{\partial^2 U}{\partial Q_y \partial Q_\eta} \tag{2.8}$$

2.3.2 Beam Equation

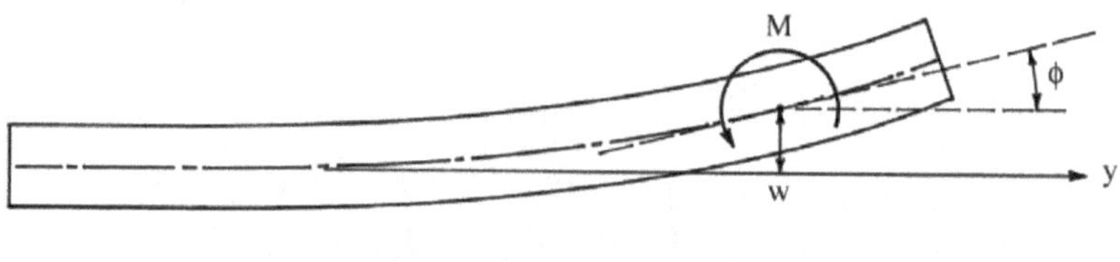

The strain energy is given by

$$U = \frac{1}{2}\int_0^L M\,d\phi$$

Also, one has that

$$\phi = \frac{dw}{dy} \rightarrow \frac{d\phi}{dy} = \frac{d^2w}{dy^2}$$

Further, beam theory gives

$$\frac{d^2w}{dy^2} = \frac{M}{EI}$$

Thus, one obtains

$$U = \frac{1}{2}\int_0^L \frac{M^2}{EI}dy = \frac{1}{2}\int_0^L EI\left(\frac{d^2w}{dy^2}\right)^2 dy$$

Now, change y to λ:

$$U = \frac{1}{2}\int_0^L \frac{M^2}{EI}d\lambda,\ M = M(\lambda)$$

Equation (2.6) gives that

$$w(y) = \frac{\partial U}{\partial Q_y} = \int_0^L \frac{m(y,\lambda)M(\lambda)}{EI}d\lambda$$

where

$$m(y,\lambda) \equiv \frac{\partial M}{\partial Q_y} = \text{ bending moment distribution due to a unit force at } y.$$

Finally, from Equation (2.8), one obtains:

$$C^{zz}(y,\eta) = \frac{\partial w(y)}{\partial Q_\eta} = \frac{\partial^2 U}{\partial Q_y \partial Q_\eta} = \int_0^L \frac{m(y,\lambda)m(\eta,\lambda)}{EI}d\lambda \tag{2.9}$$

where

$$m(\eta,\lambda) \equiv \frac{\partial M}{\partial Q_\eta} = \text{ bending moment distribution due to a unit force at } \eta.$$

An alternative derivation may be obtained from *Aeroelasticity* by R. L. Bisplinghoff, H. Ashley, and R. L. Halfman, Addison-Wesley, 1957, pp. 37–38:

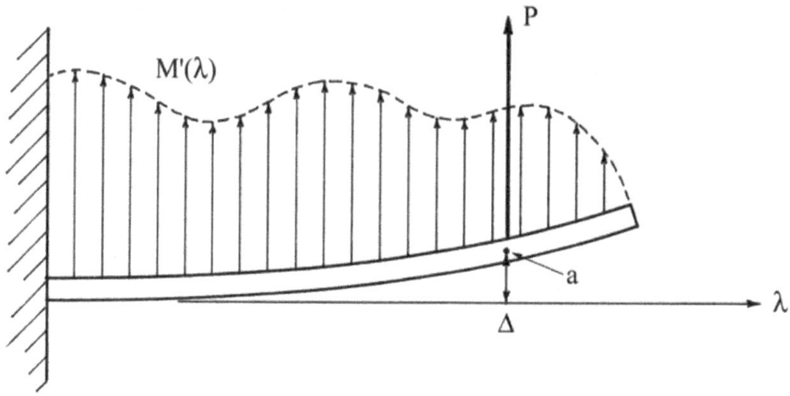

Let $M(\lambda)=M'(\lambda)+mP$, where m is the moment distribution due to a unit load at point a. Therefore, the strain energy is given by:

$$U=\int_0^L \frac{(M'+mP)^2}{2EI}d\lambda$$

The deflection at point a due to M' and P is obtained from Castigliano's first theorem as:

$$\Delta=\frac{\partial U}{\partial P}=\int_0^L \frac{(M'+mP)m}{EI}d\lambda$$

Further, the deflection at point a, due to M' alone, is found by setting $P=0$:

$$\Delta=\int_0^L \frac{M'm}{EI}d\lambda$$

Therefore, one may obtain the deflection at point y due to a general moment distribution, $M(y)$:

$$w(y)=\int_0^L \frac{M(\lambda)m(y,\lambda)}{EI}d\lambda$$

where

$$m(y,\lambda)\equiv\frac{\partial M}{\partial Q_y}=\text{ bending moment distribution due to a unit force at } y.$$

Finally, the flexibility influence coefficient function is given by

$$C^{zz}(y,\eta)=\frac{\partial w(y)}{\partial Q_\eta}=\frac{\partial^2 U}{\partial Q_y\,\partial Q_\eta}=\int_0^L \frac{m(y,\lambda)m(\eta,\lambda)}{EI}d\lambda$$

where

$$m(\eta,\lambda) \equiv \frac{\partial M}{\partial Q_\eta} = \text{ bending moment distribution due to a unit force at } \eta.$$

which is the same result as given by Equation (2.9). Note that one may extend the definitions of the moments to obtain functions for $C^{z\phi}$, $C^{\phi z}$, and $C^{\phi\phi}$.

The limits of integration have been expressed as from the base to the end of the beam (from 0 to L). However, the upper limit of the integral may be the lesser value of y or η because the corresponding cantilever beam m functions are only non-zero between the base and the y or η locations of the unit forces:

$$m(y,\lambda) = 1\times(y-\lambda) \text{ if } \lambda \le y;\ m(y,\lambda) = 0 \text{ if } \lambda \ge y$$
$$m(\eta,\lambda) = 1\times(\eta-\lambda) \text{ if } \lambda \le \eta;\ m(\eta,\lambda) = 0 \text{ if } \lambda \ge \eta$$

2.3.2.1 Example 2.3

Consider a uniform cantilevered beam with a unit force located at $L/2$:

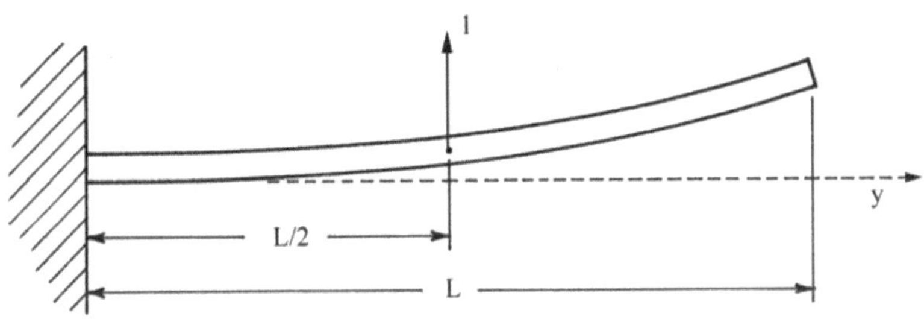

$$C^{zz}(L,L/2) = \frac{1}{EI}\int_0^{\min(y \text{ or } \eta)} (L-\lambda)(L/2-\lambda)\,d\lambda =$$

$$C^{zz}(L,L/2) = \frac{1}{EI}\int_0^{L/2} (L-\lambda)(L/2-\lambda)\,d\lambda = \frac{5}{48}\frac{L^3}{EI}$$

2.3.3 Torsion Member Equation

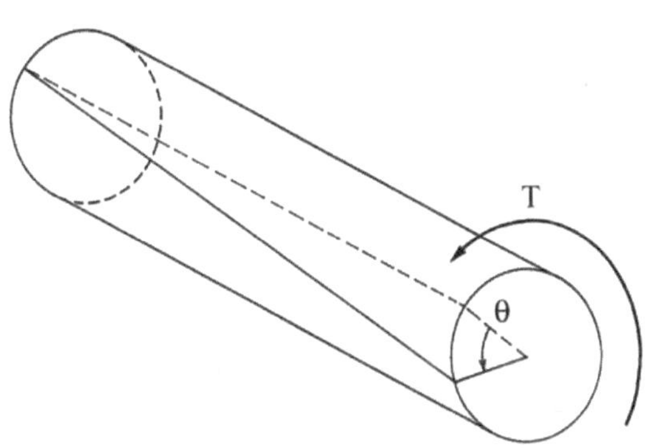

The strain energy is given by

$$U = \frac{1}{2}\int_0^L T\,d\theta$$

Torsion member theory gives that

$$\frac{d\theta}{dy} = \frac{T}{GJ}$$

from which one obtains that

$$U = \frac{1}{2}\int_0^L \frac{T^2}{GJ}dy = \frac{1}{2}\int_0^L GJ\left(\frac{d\theta}{dy}\right)^2 dy$$

By following the same procedure as for the beam derivation, one may obtain an expression for the torsional flexibility influence coefficient:

$$C^{\theta\theta}(y,\eta) = \int_0^L \frac{t(y,\lambda)t(\eta,\lambda)}{GJ}d\lambda$$

where

$$t(y,\lambda) = \frac{\partial T}{\partial Q_y} \text{ and } t(\eta,\lambda) = \frac{\partial T}{\partial Q_\eta}$$

As for the beam, the limit of integration for the cantilevered torsion member is only up to the lower value of y or η:

$$t(y,\lambda) = 1 \text{ if } \lambda \le y;\ t(y,\lambda) = 0 \text{ if } \lambda \ge y$$

$$t(\eta,\lambda) = 1 \text{ if } \lambda \le \eta;\ t(\eta,\lambda) = 0 \text{ if } \lambda \ge \eta$$

In summary, one has that

$$C^{\theta\theta}(y,\eta) = \int_0^{\min(y \text{ or } \eta)} \frac{t(y,\lambda)t(\eta,\lambda)}{GJ}d\lambda = \int_0^{\min(y \text{ or } \eta)} \frac{d\lambda}{GJ}$$

2.3.3.1 Example 2.4

Consider a finite wing as shown as follows, drawn from *Aeroelasticity* by R. L. Bisplinghoff, H. Ashley, and R. L. Halfman, Addison-Wesley, 1957, p. 45:

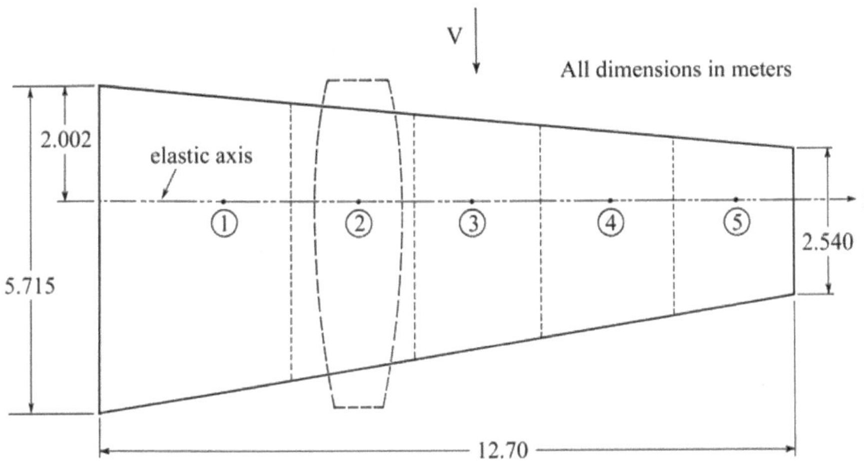

This may be structurally approximated by a beam with discrete segments:

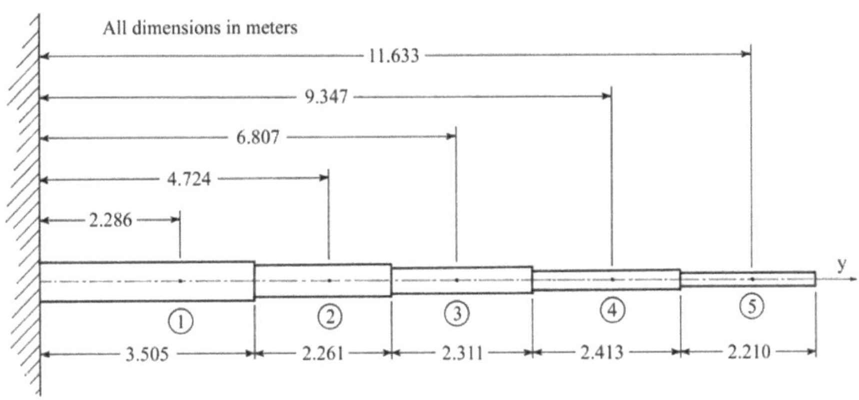

for which the values of GJ and EI are:

Segment Number	$EI\left(N-m^2\right)$	$GJ\left(N-m^2/rad\right)$
1	12.512×10^7	7.716×10^7
2	7.576×10^7	7.716×10^7
3	4.728×10^7	6.523×10^7
4	2.519×10^7	3.242×10^7
5	1.548×10^7	1.396×10^7

The flexibility influence coefficients for bending may be found from

$$C_{ij}^{zz} = \int_0^{\min\left(y_i \text{ or } y_j\right)} \frac{\left(y_i - \lambda\right)\left(y_j - \lambda\right)}{EI} d\lambda$$

which, for this discrete element case, becomes:

$$C_{ij}^{zz} = y_i y_j \int_0^{\min(y_i \text{ or } y_j)} \frac{d\lambda}{EI} - (y_i + y_j) \int_0^{\min(y_i \text{ or } y_j)} \frac{\lambda d\lambda}{EI} + \int_0^{\min(y_i \text{ or } y_j)} \frac{\lambda^2 d\lambda}{EI}$$

Therefore, one would obtain $C_{3,2}^{zz}$ as follows:

$$C_{3,2}^{zz} = 6.807 \times 4.724 \times I_1 - (6.807 + 4.724) \times I_2 + I_3$$

where

$$I_1 = \frac{1}{(EI)_1} \int_0^{3.505} d\lambda + \frac{1}{(EI)_2} \int_{3.505}^{4.724} d\lambda,$$

$$I_2 = \frac{1}{(EI)_1} \int_0^{3.505} \lambda d\lambda + \frac{1}{(EI)_2} \int_{3.505}^{4.724} \lambda d\lambda$$

$$I_3 = \frac{1}{(EI)_1} \int_0^{3.505} \lambda^2 d\lambda + \frac{1}{(EI)_2} \int_{3.505}^{4.724} \lambda^2 d\lambda$$

This gives the result

$$C_{3,2}^{zz} = 4.7778 \times 10^{-7} \ m/N$$

In a similar fashion, the complete bending flexibility influence coefficient matrix is obtained:

$$\left[C^{zz}\right] = \begin{bmatrix} 0.318 & 0.827 & 1.262 & 1.793 & 2.270 \\ 0.827 & 2.840 & 4.778 & 7.141 & 9.268 \\ 1.262 & 4.778 & 9.060 & 14.594 & 19.575 \\ 1.793 & 7.141 & 14.594 & 26.558 & 38.077 \\ 2.270 & 9.268 & 19.575 & 38.077 & 59.560 \end{bmatrix} \times 10^{-7} \ (m/N)$$

The torsional flexibility influence coefficients may be found in a similar fashion. In this case, the governing equation is

$$C_{ij}^{\theta\theta} = \int_0^{\min(y_i \text{ or } y_j)} \frac{d\lambda}{GJ}$$

so that, for example, $C_{3,2}^{\theta\theta}$ is given by

$$C_{3,2}^{\theta\theta} = \frac{1}{(GJ)_1} \int_0^{3.505} d\lambda + \frac{1}{(GJ)_2} \int_{3.505}^{4.724} d\lambda = 6.122 \times 10^{-8} \ rad/(N-m)$$

The complete torsional flexibility influence coefficient matrix is:

$$\left[C^{\theta\theta}\right]=\begin{bmatrix}2.96 & 2.96 & 2.96 & 2.96 & 2.96\\ 2.96 & 6.12 & 6.12 & 6.12 & 6.12\\ 2.96 & 6.12 & 9.07 & 9.07 & 9.07\\ 2.96 & 6.12 & 9.07 & 14.93 & 14.93\\ 2.96 & 6.12 & 9.07 & 14.93 & 26.65\end{bmatrix}\times 10^{-8}\left(rad/(N-m)\right)$$

2.3.4 Swept Beam Equations

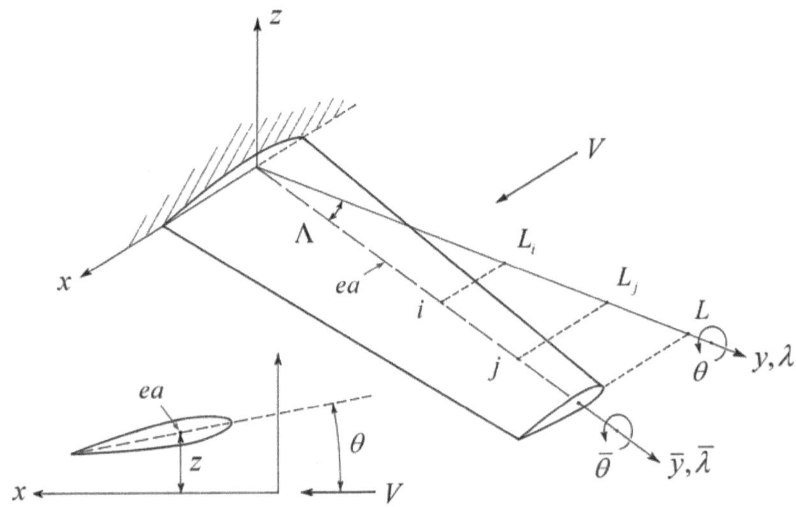

Consider a wing with a uniformly swept elastic axis, as shown earlier. A vertical displacement at the i location, z_i, is given by a discrete-element version of the influence function equation (2.4):

$$z_i=\sum_{j=1}^{n}C_{ij}^{zz}\left(Q_z\right)_j+\sum_{j=1}^{n}C_{ij}^{z\theta}\left(Q_\theta\right)_j$$

where Q_z is the applied force in the z direction and Q_θ is the applied moment about the y axis (in the θ direction). Also, from Castigliano's theorem, z_i is given by

$$z_i=\frac{\partial U}{\partial\left(Q_z\right)_i}$$

Therefore, these equations combine to give:

$$C_{ij}^{zz}=\frac{\partial z_i}{\partial\left(Q_z\right)_j}=\frac{\partial^2 U}{\partial\left(Q_z\right)_i\partial\left(Q_z\right)_j},\ C_{ij}^{z\theta}=\frac{\partial z_i}{\partial\left(Q_\theta\right)_j}=\frac{\partial^2 U}{\partial\left(Q_z\right)_i\partial\left(Q_\theta\right)_j}\tag{2.10}$$

Further, the torsional rotation about the y axis at station i, θ_i, is similarly obtained from Equation (2.5) and Castigliano's theorem:

$$\theta_i = \sum_{j=1}^{n} C_{ij}^{\theta z} (Q_z)_j + \sum_{j=1}^{n} C_{ij}^{\theta\theta} (Q_\theta)_j = \frac{\partial U}{\partial (Q_\theta)_i}$$

so that one obtains

$$C_{ij}^{\theta z} = \frac{\partial \theta_i}{\partial (Q_z)_j} = \frac{\partial^2 U}{\partial (Q_\theta)_i \partial (Q_z)_j}, C_{ij}^{\theta\theta} = \frac{\partial \theta_i}{\partial (Q_\theta)_j} = \frac{\partial^2 U}{\partial (Q_\theta)_i \partial (Q_\theta)_j} \tag{2.11}$$

Note that

$$C_{ij}^{z\theta} = C_{ji}^{\theta z}$$

as one would expect.

Now, the strain energy is due to bending and twisting along the elastic axis:

$$U = U_{bending} + U_{twisting}$$

where

$$U_{bending} = \frac{1}{2}\int_0^{\overline{L}} \frac{\overline{M}^2}{EI} d\overline{\lambda}, U_{twisting} = \frac{1}{2}\int_0^{\overline{L}} \frac{\overline{T}^2}{GJ} d\overline{\lambda}$$

Therefore, one has that

$$U = \frac{1}{2}\int_0^{\overline{L}} \left(\frac{\overline{M}^2}{EI} + \frac{\overline{T}^2}{GJ} \right) d\overline{\lambda}$$

where

$$\overline{M} = -Q_\theta \sin\Lambda + Q_z (\overline{y} - \overline{\lambda}),\ \overline{T} = Q_\theta \cos\Lambda$$

As previously shown with Equations (2.7) and (2.8), Equations (2.10) and (2.11) may be extended to give flexibility influence functions. For example, in terms of the $\overline{y}$ and $\overline{\eta}$ coordinates, one obtains

$$C_{ij}^{zz} = \frac{\partial^2 U}{\partial (Q_z)_i \partial (Q_z)_j} \rightarrow C^{zz}(\overline{y}, \overline{\eta}) = \frac{\partial^2 U}{\partial (Q_z)_{\overline{y}} \partial (Q_z)_{\overline{\eta}}}$$

\which is evaluated as follows:

$$\frac{\partial U}{\partial (Q_z)_{\overline{y}}} = \int_0^{\overline{L}} \left(\frac{\overline{M}}{EI} \frac{\partial \overline{M}}{\partial (Q_z)_{\overline{y}}} + \frac{\overline{T}}{GJ} \frac{\partial \overline{T}}{\partial (Q_z)_{\overline{y}}} \right) d\overline{\lambda} = \int_0^{\overline{L}} \frac{\overline{M}}{EI} (\overline{y} - \overline{\lambda}) d\overline{\lambda}$$

Next,

$$\frac{\partial}{\partial (Q_z)_{\bar{\eta}}}\left(\frac{\partial U}{\partial (Q_z)_{\bar{y}}}\right)=\int_0^{\bar{L}}\frac{(\bar{\eta}-\bar{\lambda})(\bar{y}-\bar{\lambda})}{EI}d\bar{\lambda}=C^{zz}(\bar{y},\bar{\eta})$$

which, in terms of discrete stations on the wing, may be written as

$$C_{ij}^{zz}=\int_0^{\min(\bar{L}_i \text{ or } \bar{L}_j)}\frac{(\bar{L}_i-\bar{\lambda})(\bar{L}_j-\bar{\lambda})}{EI}d\bar{\lambda}$$

In a similar fashion, one may obtain:

$$C_{ij}^{z\theta}=-\sin\Lambda\int_0^{\min(\bar{L}_i \text{ or } \bar{L}_j)}\frac{(\bar{L}_i-\bar{\lambda})}{EI}d\bar{\lambda}$$

$$C_{ij,}^{\theta\theta}=\cos^2\Lambda\int_0^{\min(\bar{L}_i \text{ or } \bar{L}_j)}\frac{d\bar{\lambda}}{GJ}+\sin^2\Lambda\int_0^{\min(\bar{L}_i \text{ or } \bar{L}_j)}\frac{d\bar{\lambda}}{EI}$$

This may be expressed in terms of the unswept coordinate system by noting that

$$\bar{L}_i,\bar{L}_j,\bar{\lambda},d\bar{\lambda}=(L_i,L_j,\lambda,d\lambda)/\cos\Lambda$$

Thus, one has

$$C_{ij}^{zz}=\frac{1}{\cos^3\Lambda}\int_0^{\min(L_i \text{ or } L_j)}\frac{(L_i-\lambda)(L_j-\lambda)}{EI}d\lambda$$

$$C_{ij}^{z\theta}=-\frac{\tan\Lambda}{\cos\Lambda}\int_0^{\min(L_i \text{ or } L_j)}\frac{(L_i-\lambda)}{EI}d\lambda$$

$$C_{ij}^{\theta\theta}=\cos\Lambda\int_0^{\min(L_i \text{ or } L_j)}\frac{d\lambda}{GJ}+\frac{\sin^2\Lambda}{\cos\Lambda}\int_0^{\min(L_i \text{ or } L_j)}\frac{d\lambda}{EI}$$

Also note that

$$C_{ji}^{\theta z}=C_{ij}^{z\theta}$$

2.3.4.1 Example 2.5

If the elastic axis of the wing in Example 2.4 is swept back 20 degrees, one may use the preceding equations to obtain the flexibility influence coefficients as:

$$\left[C^{z\theta}\right]=-\begin{bmatrix} 7.1 & 7.1 & 7.1 & 7.1 & 7.1 \\ 22.4 & 31.8 & 31.8 & 31.8 & 31.8 \\ 35.4 & 63.2 & 74.5 & 74.5 & 74.5 \\ 51.3 & 101.6 & 143.9 & 172.4 & 172.4 \\ 65.6 & 136.0 & 205.4 & 295.2 & 336.3 \end{bmatrix}\times 10^{-9}\left(m/(N-m)\right)$$

$$\left[C^{\theta z}\right]=-\begin{bmatrix} 7.1 & 22.4 & 35.4 & 51.3 & 65.6 \\ 7.1 & 31.8 & 63.2 & 101.6 & 136.0 \\ 7.1 & 31.8 & 74.5 & 143.9 & 206.4 \\ 7.1 & 31.8 & 74.5 & 172.4 & 295.2 \\ 7.1 & 31.8 & 74.5 & 172.4 & 336.3 \end{bmatrix}\times 10^{-9}\left(rad/N\right)$$

$$\left[C^{\theta\theta}\right]=\begin{bmatrix} 2.83 & 2.83 & 2.83 & 2.83 & 2.83 \\ 2.83 & 5.92 & 5.92 & 5.92 & 5.92 \\ 2.83 & 5.92 & 8.94 & 8.94 & 8.94 \\ 2.83 & 5.92 & 8.94 & 15.02 & 15.02 \\ 2.83 & 5.92 & 8.94 & 15.02 & 26.76 \end{bmatrix}\times 10^{-8}\left(rad/(N-m)\right)$$

The C^{zz} matrix remains the same as for the unswept case.

2.3.4.2 Example 2.6

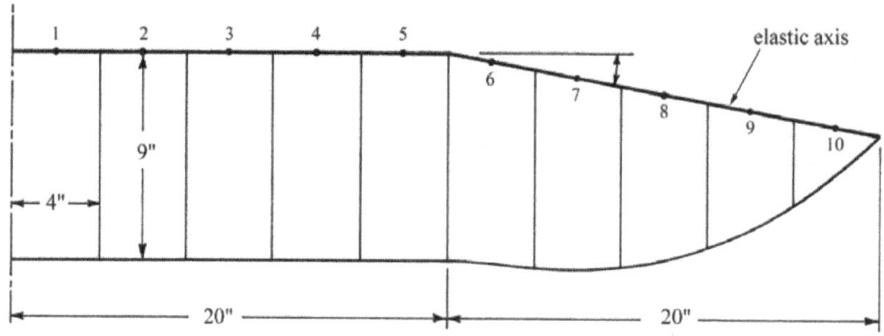

A wing for a model ornithopter (flapping-wing airplane) is shown in the preceding figure. This is characterized by having the elastic axis along the leading edge, well forward of the aerodynamic centers of pressure, so that the aerodynamic forces produced by the flapping provide a twisting motion to the structure. This twisting action is important for the efficient operation of an ornithopter wing, which contrasts with airplane wings that are usually designed to have minimum aeroelastic twist. Specific stations are chosen to which the influence coefficients are applied, and the elastic parameters are:

Station Number	$EI\left(lb - ft^2\right)$	$GJ\left(lb - ft^2/rad\right)$
1	9.867	1.292
2	7.934	1.033
3	6.286	0.814
4	4.896	0.63
5	3.738	0.478
6	2.788	0.357
7	2.023	0.257
8	1.419	0.179
9	0.955	0.12
10	0.61	0.076

From this, and the geometry of the elastic axis, the following C^{zz} flexibility influence coefficient values are obtained:

$$[C^{zz}]=\begin{bmatrix} 0 & 0.001 & 0.001 & 0.002 & 0.002 & 0.003 & 0.003 & 0.003 & 0.004 & 0.004 \\ 0.001 & 0.005 & 0.009 & 0.014 & 0.019 & 0.023 & 0.028 & 0.033 & 0.037 & 0.042 \\ 0.001 & 0.009 & 0.022 & 0.037 & 0.051 & 0.065 & 0.079 & 0.093 & 0.107 & 0.122 \\ 0.002 & 0.014 & 0.039 & 0.066 & 0.096 & 0.126 & 0.157 & 0.187 & 0.218 & 0.248 \\ 0.002 & 0.019 & 0.051 & 0.096 & 0.150 & 0.205 & 0.260 & 0.316 & 0.371 & 0.427 \\ 0.003 & 0.023 & 0.065 & 0.126 & 0.205 & 0.302 & 0.402 & 0.501 & 0.600 & 0.700 \\ 0.003 & 0.028 & 0.079 & 0.157 & 0.260 & 0.402 & 0.558 & 0.719 & 0.879 & 1.039 \\ 0.003 & 0.033 & 0.093 & 0.187 & 0.316 & 0.501 & 0.719 & 0.958 & 1.202 & 1.446 \\ 0.004 & 0.037 & 0.107 & 0.218 & 0.371 & 0.600 & 0.879 & 1.202 & 1.557 & 1.918 \\ 0.004 & 0.042 & 0.122 & 0.248 & 0.427 & 0.700 & 1.039 & 1.446 & 1.918 & 2.439 \end{bmatrix}$$

These are given in dimensions of *ft/lb*. Also, note how the largest value is $C_{10,10}^{zz}$, as one would expect from a cantilevered beam.

Next, the $C^{\theta z}$ influence coefficients are:

$$[C^{\theta z}]=-\begin{bmatrix} 0 & 0 & 0 & 0 & 0 & 0.008 & 0.015 & 0.023 & 0.030 & 0.038 \\ 0 & 0 & 0 & 0 & 0 & 0.027 & 0.053 & 0.080 & 0.106 & 0.133 \\ 0 & 0 & 0 & 0 & 0 & 0.051 & 0.101 & 0.152 & 0.202 & 0.253 \\ 0 & 0 & 0 & 0 & 0 & 0.082 & 0.163 & 0.245 & 0.327 & 0.409 \\ 0 & 0 & 0 & 0 & 0 & 0.123 & 0.245 & 0.368 & 0.491 & 0.613 \\ 0 & 0 & 0 & 0 & 0 & 0.126 & 0.256 & 0.386 & 0.516 & 0.646 \\ 0 & 0 & 0 & 0 & 0 & 0.126 & 0.261 & 0.401 & 0.540 & 0.680 \\ 0 & 0 & 0 & 0 & 0 & 0.126 & 0.261 & 0.408 & 0.561 & 0.715 \\ 0 & 0 & 0 & 0 & 0 & 0.126 & 0.261 & 0.408 & 0.572 & 0.746 \\ 0 & 0 & 0 & 0 & 0 & 0.126 & 0.261 & 0.408 & 0.572 & 0763 \end{bmatrix}$$

These are given in dimensions of *radian/lb.* Note the zero portion of the matrix. This states that forces applied to stations 1–5 (the unswept part of the elastic axis) do not give any twisting anywhere on the wing, as one would expect. However, forces applied to the outer swept part do twist the wing, which is why sweep is an important feature for the ornithopter wing.

Other cross terms are the $C^{z\theta}$ influence coefficients:

$$[C^{z\theta}] = -\begin{bmatrix} 0 & 0 & 0 & 0 & 0 & 0 & 0 & 0 & 0 & 0 \\ 0 & 0 & 0 & 0 & 0 & 0 & 0 & 0 & 0 & 0 \\ 0 & 0 & 0 & 0 & 0 & 0 & 0 & 0 & 0 & 0 \\ 0 & 0 & 0 & 0 & 0 & 0 & 0 & 0 & 0 & 0 \\ 0 & 0 & 0 & 0 & 0 & 0 & 0 & 0 & 0 & 0 \\ 0.008 & 0.027 & 0.051 & 0.082 & 0.123 & 0.126 & 0.126 & 0.126 & 0.126 & 0.126 \\ 0.015 & 0.053 & 0.101 & 0.163 & 0.245 & 0.256 & 0.261 & 0.261 & 0.261 & 0.261 \\ 0.023 & 0.080 & 0.152 & 0.245 & 0.268 & 0.386 & 0.401 & 0.408 & 0.408 & 0.408 \\ 0.030 & 0.106 & 0.202 & 0.327 & 0.491 & 0.516 & 0.540 & 0.561 & 0.572 & 0.572 \\ 0.038 & 0.133 & 0.253 & 0.409 & 0.613 & 0.646 & 0.680 & 0.715 & 0.746 & 0.763 \end{bmatrix}$$

These have the dimensions of *ft/(ft-lb).* Also, the zero portion means that torsion loads applied to the wing produce no vertical displacement in the straight unswept part of the elastic axis. Note the transpose symmetry between $C^{\theta z}$ and $C^{z\theta}$, as defined earlier.

Finally, the $C^{\theta\theta}$ influence coefficients are:

$$\left[C^{\theta\theta}\right] = \begin{bmatrix} 0.129 & 0.129 & 0.129 & 0.129 & 0.129 & 0.129 & 0.129 & 0.129 & 0.129 & 0.129 \\ 0.129 & 0.452 & 0.452 & 0.452 & 0.452 & 0.452 & 0.452 & 0.452 & 0.452 & 0.452 \\ 0.129 & 0.452 & 0.861 & 0.861 & 0.861 & 0.861 & 0.861 & 0.861 & 0.861 & 0.861 \\ 0.129 & 0.452 & 0.861 & 1.390 & 1.390 & 1.390 & 1.390 & 1.390 & 1.390 & 1.390 \\ 0.129 & 0.452 & 0.861 & 1.390 & 2.088 & 2.088 & 2.088 & 2.088 & 2.088 & 2.088 \\ 0.129 & 0.452 & 0.861 & 1.390 & 2.088 & 3.011 & 3.011 & 3.011 & 3.011 & 3.011 \\ 0.129 & 0.452 & 0.861 & 1.390 & 2.088 & 3.011 & 4.293 & 4.293 & 4.293 & 4.293 \\ 0.129 & 0.452 & 0.861 & 1.390 & 2.088 & 3.011 & 4.293 & 6.134 & 6.134 & 6.134 \\ 0.129 & 0.452 & 0.861 & 1.390 & 2.088 & 3.011 & 4.293 & 6.134 & 8.880 & 8.880 \\ 0.129 & 0.452 & 0.861 & 1.390 & 2.088 & 3.011 & 4.293 & 6.134 & 8.880 & 13.216 \end{bmatrix}$$

The dimensions are *radians/(ft-lb).* Notice the sequence of values. That is, $C_{1,1}^{\theta\theta}$ is the twist at station 1 due to a unit torsion at that station. Since $C_{2,1}^{\theta\theta}$, etc., are the twist values at other stations along the axis due to the same unit torsion at station 1, it makes sense that these are identical. When the torsion load is applied to station 2, the twist value at that station is larger, $C_{2,2}^{\theta\theta}$, because the portion of elastic axis between the base and station 2 is larger. However, the subsequent values, $C_{3,2}^{\theta\theta}$, etc., are also identical. The largest twist value is, of course, that due to a unit torsion at the end, $C_{10,10}^{\theta\theta}$.

2.3.5 Differential Equation Method for Finding Influence Functions

Influence functions may also be obtained by a direct integration of the fundamental differential equations for the beam and torsion member:

$$\text{Beam: } \frac{d^2w}{dy^2} = \frac{M}{EI}, \text{ Torsion Member: } \frac{d\theta}{dy} = \frac{T}{GJ}$$

Consider determining $C^{zz}(L/2, L)$ for a uniform cantilever beam. The moment distribution for a unit force applied at L is given by:

$$M(y) = 1 \times (L - y)$$

Therefore, one has

$$\frac{d^2w}{dy^2} = \frac{(L-y)}{EI}$$

Integrating this gives

$$\frac{dw}{dy} = \frac{1}{EI}\left(Ly - \frac{y^2}{2}\right) + C_1$$

However,

$$\frac{dw}{dy} = 0 \text{ at } y = 0, \text{ so } C_1 = 0$$

Upon integration again, one obtains

$$w = \frac{1}{EI}\left(\frac{Ly^2}{2} - \frac{y^3}{6}\right) + C_2$$

but

$$w = 0 \text{ at } y = 0 \rightarrow C_2 = 0$$

which gives the result

$$C^{zz}(L/2, L) = \frac{5}{48}\frac{L^3}{EI}$$

Note how this is the same value as the $C^{zz}(L, L/2)$ obtained previously.

3 Divergence

3.1 REPRESENTATIVE SECTION

The concept of aeroelastic divergence will be introduced by considering an idealized two-dimensional airfoil. However, it will be subsequently found that with the appropriate choice of parameters, this can likewise represent finite-wing divergence (hence the name "representative section").

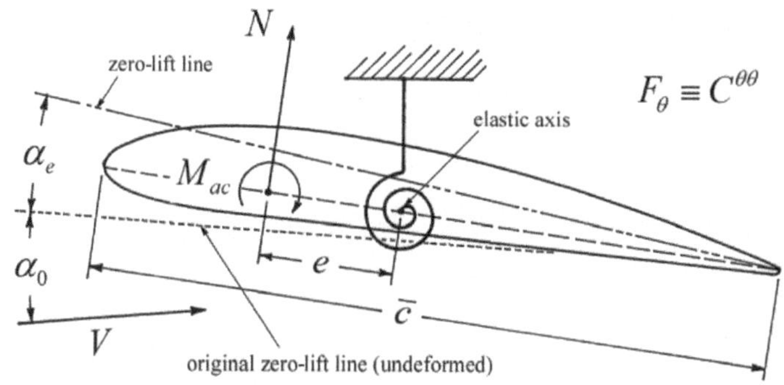

$\alpha_0 \equiv$ angle of the zero-lift line at $V = 0$

$\alpha_e \equiv$ the additional angle of the zero-lift line under aerodynamic load
$\alpha = \alpha_0 + \alpha_e \equiv$ the total angle of attack

The applied torque about the elastic axis is

$$T = Ne + M_{ac} = qS\left[C_{N\alpha}\left(\alpha_0 + \alpha_e\right)e + C_{M_{ac}}\bar{c}\right]$$

where

$$q = \text{dynamic pressure}; \quad S = \text{wing area}$$

$C_{N_\alpha} = dC_N / d\alpha$, the variation of normal-force coefficient with angle of attack
The elastic deformation is given by

$$\alpha_e = F_\theta T = F_\theta qS\left[C_{N\alpha}\left(\alpha_0 + \alpha_e\right)e + C_{M_{ac}}\bar{c}\right]$$

One may solve for α_e and obtain

$$\alpha_e = \frac{F_\theta qS\left[C_{N\alpha}\alpha_0 e + C_{M_{ac}}\bar{c}\right]}{1 - F_\theta qSC_{N_\alpha}e} \tag{3.1}$$

DOI: 10.1201/9781032709093-3

which is the change in angle of attack due to the elasticity of the wing. Note that $\alpha_e \to \infty$ when the denominator $\to 0$. Therefore, the divergence dynamic pressure is given by

$$q_{div} = \frac{1}{S C_{N_\alpha} e F_\theta} \tag{3.2}$$

This may be rearranged:

$$F_\theta S = \frac{1}{q_{div} C_{N_\alpha} e} \tag{3.3}$$

Thus, Equation (3.3) into (3.1) gives

$$\frac{\alpha_e}{\alpha_0} = \left(1 + \frac{C_{M_{ac}} \bar{c}}{C_{N_\alpha} e \alpha_0}\right) \frac{q / q_{div}}{(1 - q / q_{div})}$$

This is proportional to the elastic normal force divided by the rigid normal force, N_e / N_0, and the variation with q / q_{div} is shown as follows:

q / q_{div}	$N_e / N_0 \propto \frac{q / q_{div}}{(1 - q / q_{div})}$
0.0	0.00
0.1	0.11
0.2	0.25
0.3	0.43

Note how the variation between q / q_{div} and N_e / N_0 is nonlinear.

3.1.1 Approximate Formula for a Finite Wing

This development may be applied to an unswept finite wing if it has a high aspect ratio and uniform properties along its span (≈ constant GJ).

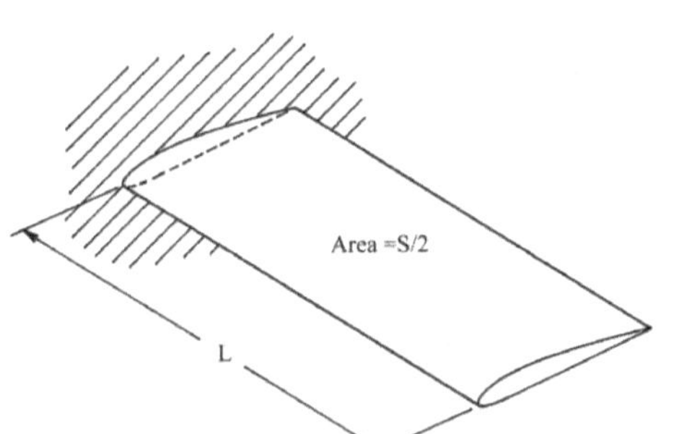

In this case, Equation (3.2) becomes

$$q_{div} \approx \frac{1}{\frac{S}{2} C_{N_\alpha} \, e F_\theta}$$

where

$$F_\theta = \int_0^{0.7L} \frac{dy}{GJ}$$

In most cases, the value of C_{N_α} is virtually the same as the wing's lift-curve slope, $C_{L\alpha}$.

3.2 DIVERGENCE OF LARGE-ASPECT-RATIO UNSWEPT WINGS

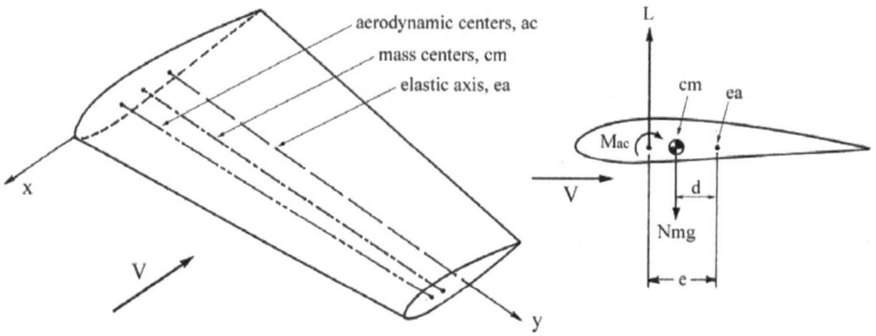

The assumptions are:

1. $\Lambda_{ea} = 0$ (no sweep)
2. Rigid chordwise segments (no chordwise deformations)
3. Beam and torsion-member model

As a consequence of 1, the torsional equilibrium case applies.

The applied torque per unit span is

$$t(y) \equiv \text{ torque per unit span about the } ea = L e + M_{ac} - N mg\, d$$

In coefficient form, this becomes

$$t(y) = q\left(ecC_L + c^2 C_{M_{ac}}\right) - N mg\, d \tag{3.4}$$

where the term $N\,mg$ is the *normal acceleration* term $\left(weight \times acceleration\ factor, N\right)$.

The differential equation for torsional equilibrium is, from before:

$$\frac{d\theta}{dy} = \frac{T}{GJ} \tag{3.5}$$

where

$$T(y) = \int_y^L t(\lambda)\, d\lambda$$

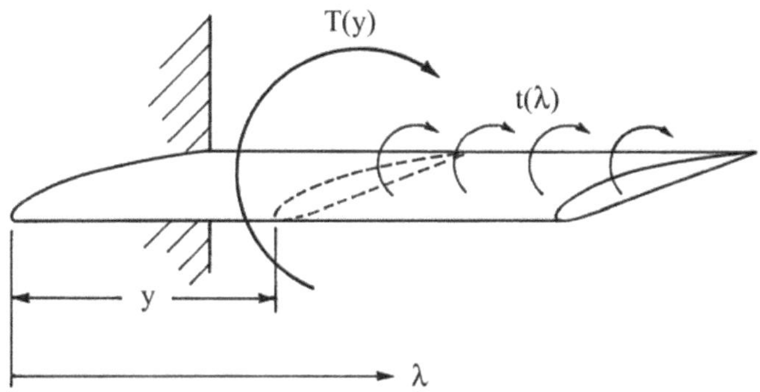

By Leibnitz's Rule (or the Fundamental Theorem of Calculus), one has

$$\frac{dT}{dy} = \int_y^L \frac{dt(y)}{d\lambda} d\lambda + t(L)\frac{dL}{dy} - t(y)\frac{dy}{dy} = 0 + 0 - t(y)\times 1 = -t(y) \tag{3.6}$$

Equations (3.4), (3.5), and (3.6) give

$$\frac{d}{dy}\left[GJ\frac{d\theta}{dy}\right] = \frac{dT}{dy} = -t(y) = -q\left(ecC_L + c^2 C_{M_{ac}}\right) + Nmg\,d \tag{3.7}$$

Now, the lift coefficient may be expressed as

$$C_L(y) = C_{L_{rigid}}(y) + C_{L_{elastic}}(y)$$

Further, $C_{L_{elastic}}$ may be written as

$$C_{L_{elastic}}(y) = a_0(y)\theta(y) \equiv a_0(y)\alpha_e(y)$$

where

$$a_0 \equiv \frac{dC_l}{d\alpha}, \text{ the sectional lift curve slope at section } y$$

Note that "strip theory" is assumed: the additional elastic aerodynamic loads on each section are due only to that section's α_e (additional elastic angle of attack).

Now, assume a special case where GJ is constant along the span, as well as e, c, m, d, and a_0. Equation (3.7) becomes

$$\frac{d^2\theta}{dy^2} + \left(\frac{eca_0}{GJ}q\right)\theta = \frac{-1}{GJ}\left(qecC_{L_{rigid}} + qc^2 C_{M_{ac}} - Nmg\,d\right) \tag{3.8}$$

Divergence is given by the solution of the homogeneous part of this:

$$\frac{d^2\theta}{dy^2} + \lambda^2\theta = 0 \text{ where } \lambda^2 = \frac{e\,c\,a_0}{GJ}q$$

This is an important concept, similar to the solutions for structural buckling, in that the condition for divergence is independent of the degree of initial loading to the system. If a solution for finite θ can be found even for zero initial loading (e.g., if the wing were at a zero-lift angle of attack), then that value of λ would yield q_{div}.

Assume a solution of the form

$$\theta = A\sin(\lambda y) + B\cos(\lambda y) \tag{3.9}$$

With the end condition of zero twist at the root, one has that $\theta(0) = 0 \rightarrow B = 0$. Also, with zero torque at the other end, Equation (3.5) gives that $\theta'(L) = 0 \rightarrow \cos(\lambda L) = 0$. Therefore, the only finite (non-trivial) values of λL that satisfy this are

$$\lambda L = (2n+1)\pi / 2 \qquad n = 0, 1, 2, 3,$$

Upon choosing the lowest value of λ, one has

$$\lambda = \frac{\pi}{2L} \rightarrow q_{div} = \frac{\pi^2}{4L^2}\frac{GJ}{cea_0} \approx \frac{2.4674}{L}\frac{GJ}{S'ea_0}; S' = Lc \tag{3.10}$$

By comparison, the $0.7\bar{L}$ approximate solution gives

$$q_{div} = \frac{1.43}{L}\frac{GJ}{S'eC_{L_\alpha}}$$

This is definitely more conservative, giving a V_{div} which is 24% less than the "exact" value.

From Equation (3.9), one obtains that the shape of the wing at divergence is given by

$$\theta_{div} = A\sin\left(\frac{\pi}{2}\frac{y}{L}\right)$$

where A is an undetermined magnitude. Note that this is *not* a uniform twist.

Equation (3.10) shows that as e increases, q_{div} decreases, whereas as J increases, q_{div} increases. That is, the further aft the elastic axis is from the aerodynamic center, the lower the speed at which divergence occurs. However, increasing the torsional stiffness moves the divergence boundary to a higher speed.

An interesting historical example of how a wing was weakened by strengthening it is given by James Edward Gordon in *Structures (or Why Things Don't Fall Down)*, Penguin Books, 1983, pp. 161–166. In this instance, a 1917 German fighter called the Fokker D-8 was being evaluated by the Army. This airplane had extraordinary promise because it was a "clean" monoplane at a time when wire-rigged biplanes were common. However, even though the wings survived the mandatory static loading tests, the Army's experts required Fokker to increase the strength of the rear spar. By so doing, the rear spar was also stiffened and the elastic axis moved aft from its original position. This then caused divergence problems that were not solved until the design went back to the original structure.

3.2.1 Integral Method, Discrete-Element Solution

In this instance, GJ, e, m, d, a_0, and c will be allowed to vary. Now, recall that $\theta(y)$ is given by

$$\theta(y)=\int_0^L C^{\theta\theta}(y,\eta)t(\eta)d\eta$$

From previous equations, this may be written as

$$\theta(y)=q\int_0^L C^{\theta\theta}(y,\eta)eca_0\,\theta\,d\eta+f(y)$$

where

$$f(y)=\int_0^L C^{\theta\theta}(y,\eta)\left(qecC_{L_r}+qc^2C_{M_{ac}}-N\,mg\,d\right)d\eta$$

Now, divide the wing into spanwise segments as shown in Figure (a):

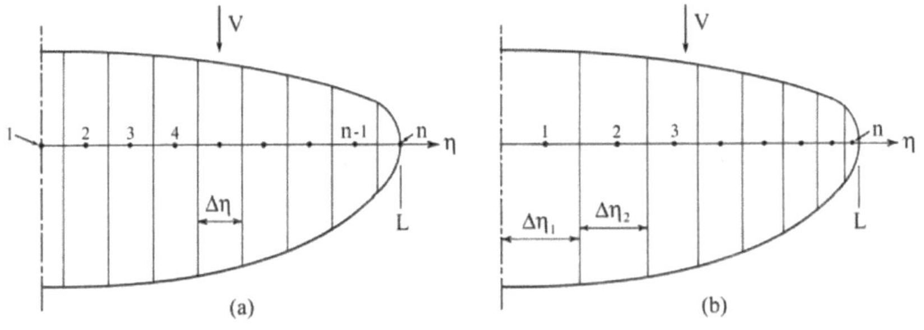

From a weighing-matrix approach, the equations become

$$\{\theta\} = q\left[C^{\theta\theta}\right][W][e][c][a_0]\{\theta\} + \{f\} \tag{3.11}$$

where

$$\{f\} = q\left[C^{\theta\theta}\right][W]\left\{ec\,C_{L_r} + c^2\,C_{M_{ac}}\right\} - N\left[C^{\theta\theta}\right][W]\{mg\,d\}$$

and

$$[W] \equiv \textit{Weighing Matrix} = \begin{bmatrix} W_1 & 0 & 0 & 0 \\ 0 & W_2 & 0 & 0 \\ . & . & . & . \\ 0 & 0 & . & W_n \end{bmatrix}$$

Example: The weighing matrix for trapezoidal integration is

$$[W]_{Trapezoidal} = \Delta\eta \begin{bmatrix} 1/2 & 0 & 0 & . & 0 & 0 & 0 & 0 \\ 0 & 1 & 0 & . & 0 & 0 & 0 & 0 \\ 0 & 0 & 1 & . & 0 & 0 & 0 & 0 \\ . & . & . & . & 0 & 0 & 0 & 0 \\ 0 & 0 & 0 & 0 & . & . & . & . \\ 0 & 0 & 0 & 0 & . & 1 & 0 & 0 \\ 0 & 0 & 0 & 0 & . & 0 & 1 & 0 \\ 0 & 0 & 0 & 0 & . & 0 & 0 & 1/2 \end{bmatrix}$$

where $\Delta\eta$ is the width of the discrete spanwise element.

Note that station 1 is at the root and station n is at the tip. Further, for a cantilevered wing, $C_{1,j}^{\theta\theta}$ and $C_{i,1}^{\theta\theta} = 0$. Also, matrices $[e]$, $[c]$, and $[a_0]$ are diagonal:

$$[e] = \begin{bmatrix} e_1 & 0 & 0 & 0 \\ 0 & e_2 & 0 & 0 \\ . & . & . & . \\ 0 & 0 & . & e_n \end{bmatrix}, \quad [c] = \begin{bmatrix} c_1 & 0 & 0 & 0 \\ 0 & c_2 & 0 & 0 \\ . & . & . & . \\ 0 & 0 & . & c_n \end{bmatrix}, \quad [a_0] = \begin{bmatrix} a_{0_1} & 0 & 0 & 0 \\ 0 & a_{0_2} & 0 & 0 \\ . & . & . & . \\ 0 & 0 & . & a_{0_n} \end{bmatrix}$$

Equation (3.11) can be written as

$$[A - \lambda I]\{\theta\} = \frac{-1}{q}\{f\} \tag{3.12}$$

where

$$\lambda = \frac{1}{q}, \quad [A] = \left[C^{\theta\theta}\right][W][e][c][a_0]$$

In this instance, the concept of divergence occurring independent of the initial loading is expressed as

$$[A-\lambda I]\{\theta\}=0$$

The eigenvalues λ_1, λ_2,……λ_n will give the q values for divergence, The lowest q is q_{div}. Also, for $q \prec q_{div}$, Equation (3.12) gives the elastic deformation shape:

$$[B]\{\theta\}=\{b\}\rightarrow\{\theta\}=[B]^{-1}\{b\}$$

where

$$[B]\equiv[A-\lambda I],\quad \{b\}\equiv\frac{-1}{q}\{f\}$$

Case (b) in the previous figure shows another discrete-element scheme in which the weighing matrix is not involved. In this case, matrices $[A]$ and $\{f\}$ are given by

$$[A]=\left[C^{\theta\theta}\right][\Delta\eta][e][c][a_0]$$

$$\{f\}=q\left[C^{\theta\theta}\right][\Delta\eta]\left\{ecC_{L_r}+c^2C_{M_{ac}}\right\}-N\left[C^{\theta\theta}\right][\Delta\eta]\{mgd\}$$

where $[\Delta\eta]$ is a diagonal matrix consisting of arbitrarily sized elements. In this case, note that station 1 is the midpoint of the first element and station n is the midpoint of the last element. Also, even for the cantilevered wing, $C_{1,j}^{\theta\theta}$ and $C_{i,1}^{\theta\theta}$ are generally ≠ 0. This is the method used for the following example.

3.2.1.1 Example 3.1

Consider the wing in Example 2.4. This has the additional characteristics:

Station Number	$a_o\left(\frac{1}{rad}\right)$	$e(m)$	$c(m)$	$\Delta\eta(m)$
1	4.742	0.514	5.144	3.505
2	4.742	0.453	4.534	2.261
3	4.742	0.401	4.013	2.311
4	4.742	0.338	3.378	2.413
5	4.742	0.281	2.807	2.210

The solution for divergence speed gives that, for atmospheric density $\rho=1.225kg/m^3$,

$$V_{div}=527m/\text{sec}.$$

This is actually higher than the assumed speed regime for this wing. The a_o and e parameters are for incompressible flow. Therefore, one may conclude that divergence will not be an operational problem for this wing design.

3.2.2 Rayleigh-Ritz Method (or Assumed-Mode Method)

Assume that $\theta(y)$ may be expressed as a sum of independent (none are a linear combination of the others) normalized shapes, $\Theta_i(y)$, each of which satisfies the end conditions:

$$\theta(y)=\sum_{i=1}^{n}\Theta_i(y)q_i \tag{3.13}$$

where q_i is the "generalized coordinate" to be determined. The end conditions are

$$\Theta_i(0)=0,\ \Theta_i'(L)=0 \ \text{(fixed at wall, no torque at } L\text{)}$$

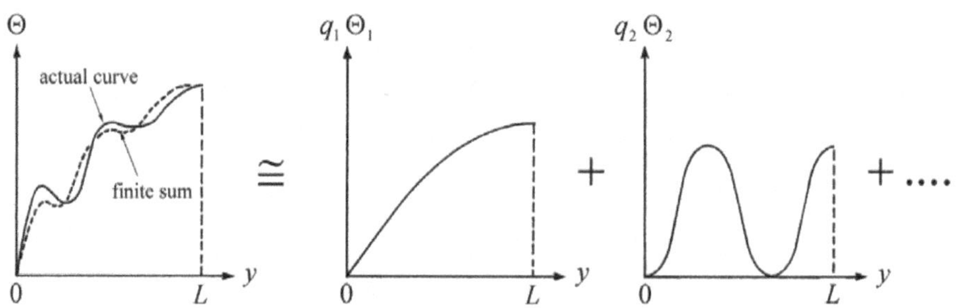

Virtual-Work Theorem: The first-order change in the total potential must vanish for every virtual displacement when an elastic body is in equilibrium:

$$\delta(U-W)=0$$

where U = strain energy, W = virtual work of the applied forces and moments. Therefore, this gives

$$\frac{\partial}{\partial q_i}(U-W)=0 \qquad i=1,2,3,\ldots\ldots,n \tag{3.14}$$

Aside: Equation (3.14) is equivalent to the static form of Lagrange's equation:

$$\frac{\partial U}{\partial q_i}=Q_i \ (\textit{generalized force})$$

From before, the strain energy for torsion is

$$U=\frac{1}{2}\int_0^L GJ\left(\frac{d\theta}{dy}\right)^2 dy$$

With Equation (3.13), this becomes

$$U=\frac{1}{2}\sum_{i=1}^{n}\sum_{j=1}^{n}\left[\int_0^L GJ\left(\frac{d\Theta_i}{dy}\right)\left(\frac{d\Theta_j}{dy}\right)dy\right]q_iq_j \tag{3.15}$$

Further, the virtual work is given by the integral of the equilibrium torque times the equilibrium twist:

$$W = \int_0^L \theta(y)t(y)dy$$

which, with Equation (3.13), becomes

$$W = \sum_{i=1}^{n}\left[\int_0^L \Theta_i(y)t(y)dy\right]q_i \tag{3.16}$$

So Equations (3.15) and (3.16) into Equation (3.14) gives:

$$\sum_{j=1}^{n}\left[\int_0^L GJ\left(\frac{d\Theta_i}{dy}\right)\left(\frac{d\Theta_j}{dy}\right)dy\right]q_j = \int_0^L \Theta_i(y)t(y)dy \quad i = 1,2,3,\ldots n \tag{3.17}$$

From previous work, the right-hand side becomes

$$\int_0^L \Theta_i\left(ecC_{L_e}q + ecC_{L_r}q + qc^2C_{M_{ac}} - Nmgd\right)dy \tag{3.18}$$

Note from Equation (3.13) that C_{L_e} may be written as

$$C_{L_e} = \left(C_{L_\alpha}\right)_e\theta(y) = \left(C_{L_\alpha}\right)_e\left(\sum_{j=1}^{n}\Theta_j q_j\right)$$

This, into Equation (3.17), gives

$$\sum_{j=1}^{n}\left[\int_0^L GJ\left(\frac{d\Theta_i}{dy}\right)\left(\frac{d\Theta_j}{dy}\right)dy - q\int_0^L \left(C_{L_\alpha}\right)_e ec\Theta_i\Theta_j\,dy\right]q_j$$
$$= \int_0^L \Theta_\iota\left(qecC_{L_r} + qc^2C_{M_{ac}} \quad Nmgd\right)dy$$

The matrix form of this is

$$[A - qB]\{q_i\} = \{C\} \tag{3.19}$$

where

$$A_{ij} = \int_0^L GJ\left(\frac{d\Theta_i}{dy}\right)\left(\frac{d\Theta_j}{dy}\right)dy, B_{ij} = \int_0^L \left(C_{L_\alpha}\right)_e ec\Theta_i\Theta_j dy$$

$$C_i = \int_0^L \Theta_i\left(qecC_{L_r} + qc^2C_{M_{ac}} - Nmgd\right)dy$$

As before, the condition for divergence is provided by the solution of the homogeneous form of Equation (3.19):

$$[A - qB]\{q_i\} = 0$$

Note that q is the dynamic pressure and q_i are the magnitudes of the assumed modes.

For $q \prec q_{div}$, the elastic deformation may be found by solving for q_i:

$$\{q_i\} = [A - qB]^{-1}\{C\}$$

and substituting these values into Equation (3.13) to obtain $\theta(y)$. Next, one may use $\theta(y)$ to calculate the increment in the bending moment and torsion loads due to elasticity.

Equation (3.19) is similar to the previous matrix formulation, Equation (3.12). However, a good choice of assumed modes, Θ_i, will greatly reduce the order of the matrix.

3.3 STATIC DIVERGENCE OF SWEEP WINGS

3.3.1 SEMI-RIGID MODEL

In the early days of aeroelastic analysis, the ability to perform numerical calculations was limited, and it was important to construct mathematical models that could be analyzed with the computational capabilities of the times. One such approach was the semi-rigid model, introduced by R. A. Frazer ("An investigation of wing flutter," ARC, R&M No. 1042, 1926), where it is assumed that the wing is a rigid panel with the elastic properties concentrated at specific hinge lines. This approach is applied to the static divergence of a swept wing, as shown in the following figure:

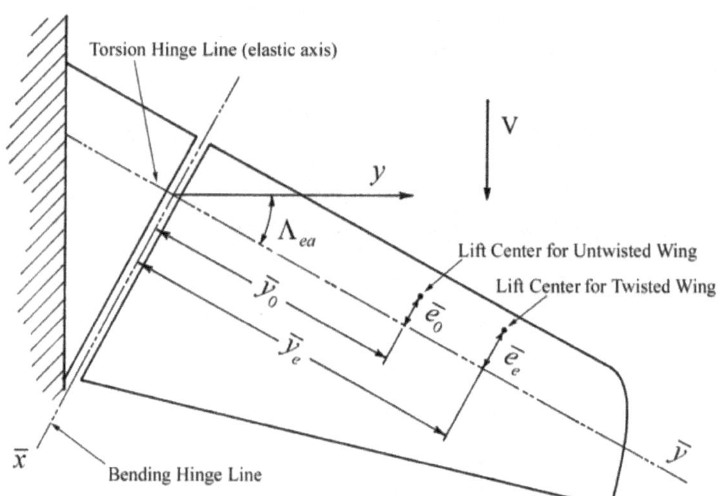

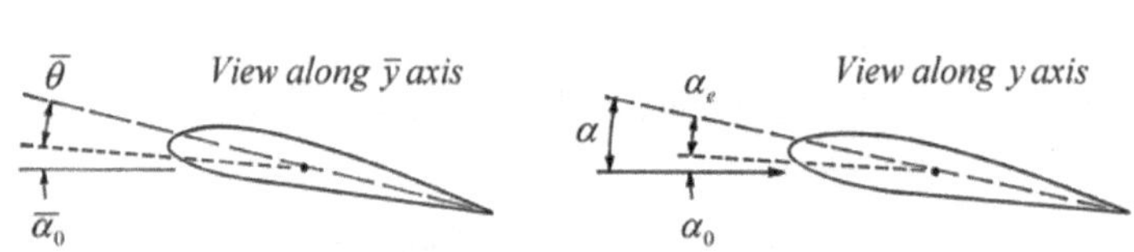

The lift of the hinged portion of the wing is given by

$$L = L_0 + L_e = q\frac{S}{2}\left(C_{L\alpha}\alpha_0 + C_{L\alpha_e}\alpha_e\right) \tag{3.20}$$

where

$C_{L\alpha}$ = the lift-curve slope of the undeflected wing panel

$C_{L\alpha_e}$ = the lift-curve slope of the deflected panel

define

$$\eta = \frac{C_{L\alpha_e}}{C_{L\alpha}}$$

so that Equation (3.20) becomes

$$L = q\frac{S}{2}(\alpha_0 + \eta\alpha_e)C_{L\alpha}$$

The moment about the bending hinge line is

$$\bar{M} = \bar{M}_0 + \bar{M}_e = q\frac{S}{2}C_{L\alpha}\left(\bar{y}_0\,\alpha_0 + \bar{y}_e\,\eta\,\alpha_e\right) \tag{3.21}$$

and the moment about the torsion hinge line is

$$\bar{T} = \bar{T}_0 + \bar{T}_e = q\frac{S}{2}C_{L\alpha}\left(\bar{e}_0\,\alpha_0 + \bar{e}_e\,\eta\,\alpha_e\right) \tag{3.22}$$

Note that α_0 is that value of α when weight and elastic deflection are in equilibrium (at $q = 0$).

The deformations due to these loads are given by

$$\bar{\theta} = F_\theta\,\bar{T}, \quad \bar{\Gamma} = F_\Gamma\,\bar{M} \tag{3.23}$$

where F_θ is the torsional spring flexibility constant and F_Γ is the bending spring flexibility constant.

The relationship between the elastic angle of attack and the deformation angles is given by

$$\alpha_e = \bar{\theta}\cos\Lambda_{ea} - \bar{\Gamma}\sin\Lambda_{ea} \tag{3.24}$$

Equations (3.21) through (3.23) into (3.24) give

$$\begin{aligned}\alpha_e = {} & \cos\Lambda_{ea}\,F_\theta\,q\frac{S}{2}C_{L\alpha}\left(\bar{e}_0\alpha_0 + \bar{e}_e\eta\alpha_e\right) \\ & - \sin\Lambda_{ea}\,F_\Gamma\,q\frac{S}{2}C_{L\alpha}\left(\bar{y}_0\alpha_0 + \bar{y}_e\eta\alpha_e\right)\end{aligned} \tag{3.25}$$

Upon solving for α_e, one obtains

$$\alpha_e = \frac{q\frac{S}{2}C_{L_\alpha}\left(F_\theta\,\overline{e}_0\,\cos\Lambda_{ea} - F_\Gamma\,\overline{y}_0\,\sin\Lambda_{ea}\right)\alpha_0}{1-q\frac{S}{2}C_{L_\alpha}\eta\left(F_\theta\,\overline{e}_e\cos\Lambda_{ea} - F_\Gamma\,\overline{y}_e\,\sin\Lambda_{ea}\right)} \tag{3.26}$$

Divergence occurs when the denominator equals zero, which gives

$$q_{div} = \frac{1}{\frac{S}{2}C_{L_\alpha}\eta\left(F_\theta\,\overline{e}_e\cos\Lambda_{ea} - F_\Gamma\,\overline{y}_e\sin\Lambda_{ea}\right)} \tag{3.27}$$

Further, one can see a condition for *no divergence* ($q_{div} = \infty$) is when the denominator of Equation (3.27) equals zero, so that

$$\Lambda_{ea} = \tan^{-1}\frac{F_\theta\,\overline{e}_e}{F_\Gamma\,\overline{y}_e}$$

A summary of these results is shown thus:

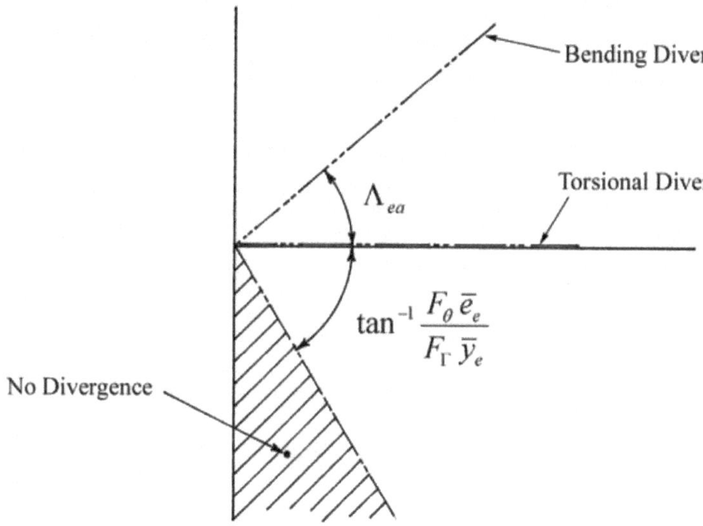

For $\Lambda_{ea} = 0$, one has torsional divergence only:

$$q_{div} = \frac{1}{\frac{S}{2}\eta C_{L_\alpha}F_\theta\,\overline{e}_e}$$

For $\Lambda_{ea} \to -90^0$, one has primarily bending divergence:

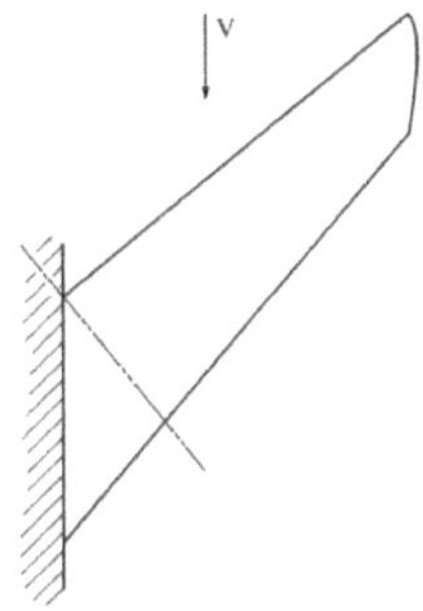

This "sweep-forward" geometry is usually a situation that aggravates the divergence problem, as seen by the following graph:

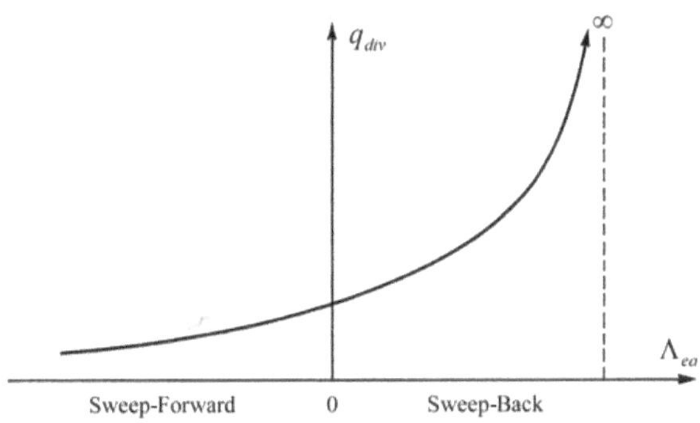

Wing sweep has been used to manage transonic flow conditions, and sweep-forward would have been preferred because it does not suffer from the tip stall problems that sweep-back causes. However, aeroelastic considerations have made sweep-back the obvious choice. An example of this is shown by the Boeing B-47 bomber of the 1950s:

This was a radical airplane in many ways. First of all, because of the requirement for flight speeds in the transonic regime, the airfoil had to be thinner than had been standard practice for previous, slower-speed bombers from Boeing. However, this thin airfoil gave rise to such flexibility that wing bending was obvious when at rest and in flight. This was rather alarming to engineers who were used to airplanes with no perceivable deformation under normal operation, and in fact, this flexibility would have caused divergence problems if the wing had not been swept back. Further, as mentioned before, the sweep-back provided an aerodynamic benefit for transonic flight. Therefore, this design solution solved two problems at once. Also, notice how the engines are placed along the wing. This distributed weight counters the spanwise lift, in a "span loading" fashion, thus reducing the bending moment at the wing root. Likewise, these are located forward relative to the wing. As explained later, this serves to suppress flutter. The significance of the B-47 was that it provided the archetype for the Boeing B-52 bomber and an entire class of civil transport aircraft, such as the Boeing 747.

Equations (3.26) and (3.27) may be combined to obtain a measure of how the loading of this wing model changes due to its elasticity:

$$\alpha_e = \frac{\eta_1}{\eta}\left(\frac{q/q_{div}}{1-q/q_{div}}\right)\alpha_0$$

where

$$\eta_1 = \frac{F_\theta\,\bar{e}_0\cos\Lambda_{ea} - F_\Gamma\,\bar{y}_0\,\sin\Lambda_{ea}}{F_\theta\,\bar{e}_e\,\cos\Lambda_{ea} - F_\Gamma\,\bar{y}_e\,\sin\Lambda_{ea}}$$

Therefore,

$$\frac{\alpha}{\alpha_0} = 1+\frac{\alpha_e}{\alpha_0} = 1+\frac{\eta_1}{\eta}\left(\frac{q/q_{div}}{1-q/q_{div}}\right) = \frac{L}{L_0}$$

The term

$$\left(\frac{q/q_{div}}{1-q/q_{div}}\right)$$

may be referred to as the "magnification factor." As well as the lift magnification, one may obtain torsional and bending moment magnification equations:

$$\frac{\bar{T}}{\bar{T}_0} = 1+\eta_1\frac{\bar{e}_e}{\bar{e}_0}\left(\frac{q/q_{div}}{1-q/q_{div}}\right) \quad \text{Torsional Magnification}$$

$$\frac{\bar{M}}{\bar{M}_0} = 1+\eta_1\frac{\bar{y}_e}{\bar{y}_0}\left(\frac{q/q_{div}}{1-q/q_{div}}\right) \quad \text{Bending Moment Magnification}$$

3.3.2 Determination of Equivalent-Spring Parameters F_Γ and F_θ

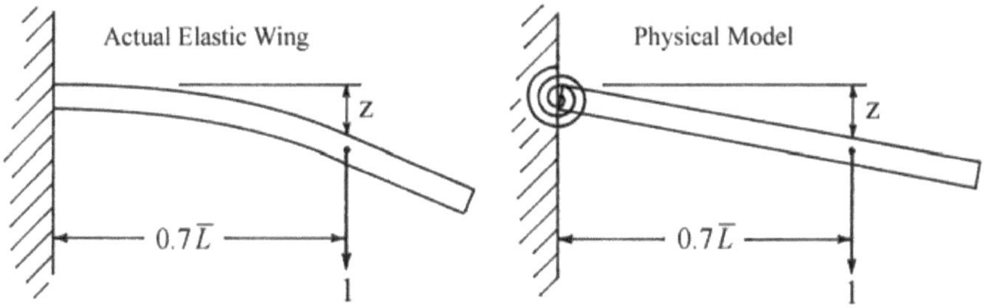

The deflection of the actual wing at $0.7\bar{L}$ due to a unit load at $0.7\bar{L}$ is given by

$$C^{zz}\left(0.7\bar{L},0.7\bar{L}\right)=\int_0^{0.7\bar{L}}\frac{\left(0.7\bar{L}-\bar{\lambda}\right)^2}{EI}d\bar{\lambda} \tag{3.28}$$

and the deflection of the physical model at $0.7\bar{L}$ is

$$z=\left(root\ angle\right)\times\left(dist.\ to\ z\right)=F_\Gamma\ \bar{M}\times 0.7\bar{L}$$

where

$$\bar{M}=0.7\bar{L}=\text{bending moment at the root due to the unit load at } 0.7\bar{L}$$

Thus, the deflection equation is

$$z=F_\Gamma\left(0.7\bar{L}\right)^2 \tag{3.29}$$

and upon equating Equations (3.28) and (3.29), one obtains

$$F_\Gamma=\frac{1}{\left(0.7\bar{L}\right)^2}\int_0^{0.7\bar{L}}\frac{\left(0.7\bar{L}-\bar{\lambda}\right)^2}{EI}d\bar{\lambda}$$

The torsion spring parameter may be derived in a similar way, which gives

$$F_\theta=\int_0^{0.7\bar{L}}\frac{d\bar{y}}{GJ}$$

3.3.3 Discrete-Element Model

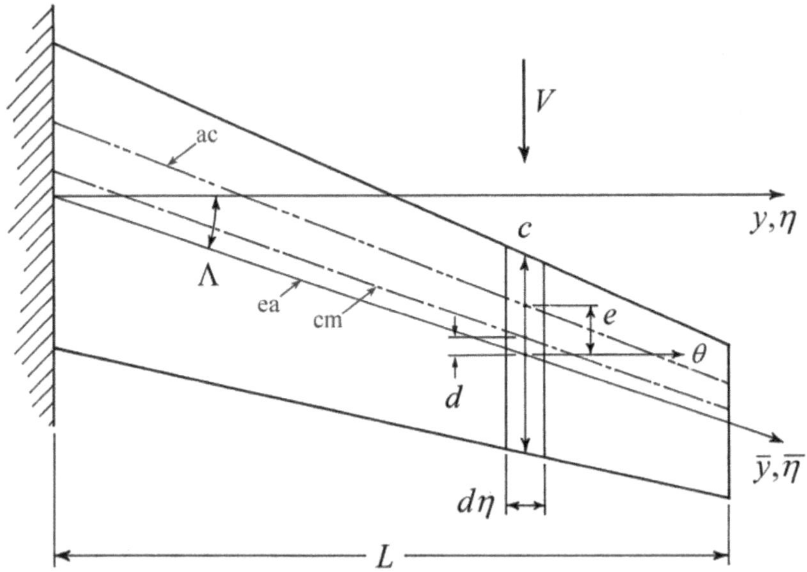

One may apply the discrete-element solution to the swept-wing problem, in the same manner as described for the unswept case. The starting point is the twist deformation equation from Section 2.2:

$$\theta(y) = \int_0^L C^{\theta z}(y,\eta) Z(\eta) d\eta + \int_0^L C^{\theta\theta}(y,\eta) t(\eta) d\eta$$

where

$$Z(\eta) = L - N\,mg = qc\left[C_{L_r}(y) + a_o(y)\theta(y)\right] - N\,mg$$

$$t(\eta) = L\,e + M_{ac} - N\,mg\,d = qce\left[C_{L_r}(y) + a_o(y)\theta(y)\right] + qc^2 C_{M_{ac}} - N\,mg\,d$$

These combine to give

$$\theta(y) = q\int_0^L C^{\theta z}(y,\eta) c a_o \theta(y) d\eta + q\int_0^L C^{\theta\theta}(y,\eta) e c a_o \theta(y) d\eta + F(y)$$

$$F(y) = \int_0^L C^{\theta z}(y,\eta)\left(qcC_{L_r} - Nmg\right) d\eta + \int_0^L C^{\theta\theta}(y,\eta)\left(qceC_{L_r} + qc^2 C_{M_{ac}} - Nmg\,d\right) d\eta$$

which may be expressed in discrete element matrix form as

$$\{\theta\} = q\left[C^{\theta z}\right][\Delta\eta][c][a_o]\{\theta\} + q\left[C^{\theta\theta}\right][\Delta\eta][e][c][a_o]\{\theta\} + \{F\}$$

$$\{F\} = q\left[C^{\theta z}\right][\Delta\eta]\left\{cC_{L_r}\right\} + N\left[C^{\theta z}\right][\Delta\eta]\{mg\}$$
$$+q\left[C^{\theta\theta}\right][\Delta\eta]\left\{ecC_{L_r} + c^2C_{M_{ac}}\right\} - N\left[C^{\theta\theta}\right][\Delta\eta]\{mg\,d\}$$

where $[\Delta\eta]$, $[c]$, $[a_o]$, and $[e]$ are diagonal square matrices as described before.
Define

$$\left[\bar{A}\right] \equiv \left[C^{\theta z}\right][\Delta\eta][c][a_o] + \left[C^{\theta\theta}\right][\Delta\eta][e][c][a_o]$$

so that

$$\{\theta\} = q\left[\bar{A}\right]\{\theta\} + \{F\} \rightarrow \left[q\bar{A} - I\right]\{\theta\} = -\{F\} \rightarrow \left[\bar{A} - \lambda I\right]\{\theta\} = -\{F\}/q$$

For divergence, one solves the eigenvalue problem:

$$\left[\bar{A} - \lambda I\right]\{\theta\} = 0, \text{ where } \lambda = 1/q$$

3.3.3.1 Example 3.2

Consider the uniform cantilevered wing shown here:

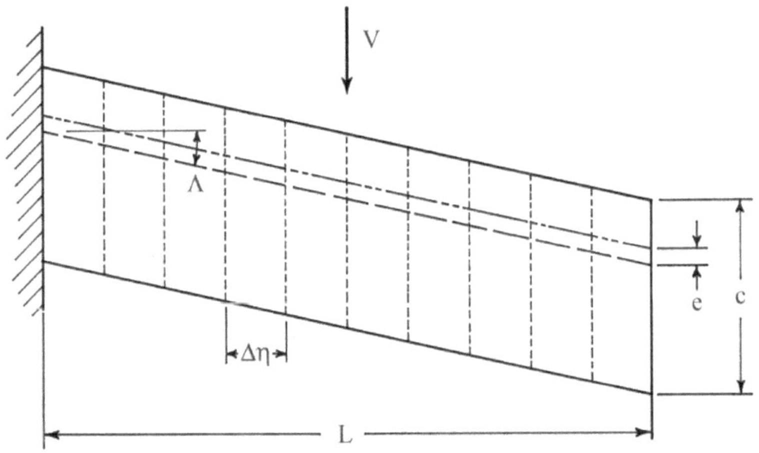

The values of e and c are uniform along the span and are given by

$$e = \bar{e}/\cos\Lambda,\ c = \bar{c}/\cos\Lambda$$

where $\bar{e} = 0.1524m$ and $\bar{c} = 1.8288m$. Also, from swept-wing aerodynamic theory,

$$a_o = \bar{a}_o \cos\Lambda$$

where, for this aspect ratio, $\bar{a}_o = 4.775/rad$. Further, the span is given by $L = \bar{L}\cos\Lambda$, where $\bar{L} = 6.096m$.

The elastic properties are also uniform along the elastic axis:

$$EI(\bar{y}) = 9.7523\times10^{6}\ N-m^{2};\ GJ(\bar{y}) = 9.8763\times10^{5}\ N-m^{2}\ /\ rad$$

from which the flexibility influence coefficient values may be calculated as described in Chapter 2. Finally, the $\Delta\eta$ increments are chosen to be uniform:

$$\Delta\eta = \Delta\bar{\eta}\cos\Lambda,\ \Delta\bar{\eta} = 0.6096m$$

The divergence speed at sea level ($\rho = 1.225kg\ /\ m^{3}$) is found for varying sweep, Λ, and the results are plotted here:

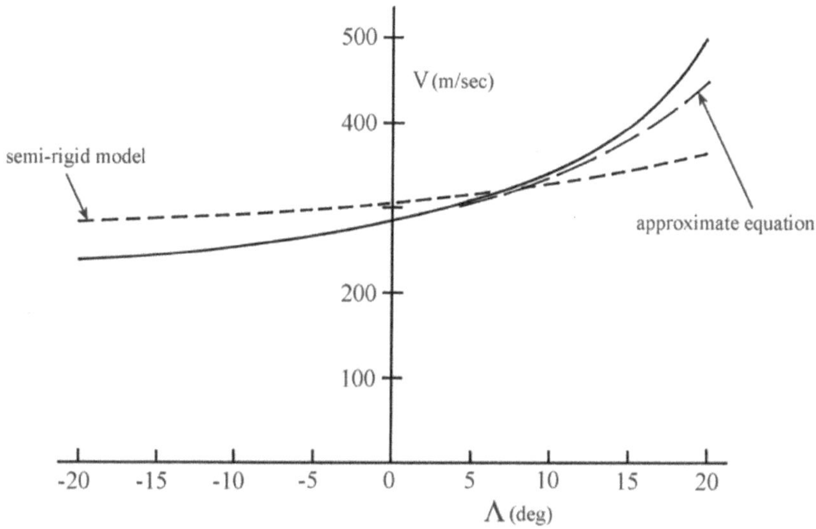

It is seen that this follows the trend of the semi-rigid case discussed before, in that sweepback suppresses divergence. However, in a quantitative sense, applying the semi-rigid equations to this example gives a generally poor comparison. Dewey H. Hodges and G. Alvin Pierce (*Introduction to Structural Dynamics and Aeroelasticity*, Cambridge, 2002, p. 102) present an alternative solution using a Ritz approximation:

$$q_{div} = \frac{\pi^{2}GJ}{4eca_{o}L^{2}\cos^{2}\Lambda\left[1-\dfrac{3\pi^{2}}{76}\dfrac{L}{e}\dfrac{GJ}{EI}\tan\Lambda\right]}$$

And the plotted results show the comparison to be excellent.

To conclude this chapter, it is important to observe that carbon fiber composite materials have allowed the "aeroelastic tailoring" of wings with extraordinary resistance to divergence. The way this is done is that the wing skin is built up of layers of unidirectional carbon in orientations such that the effective elastic axis is far forward of what would be the case if isotropic materials (such as aluminum sheet) were used. An example of this is the X-29 experimental airplane:

Dryden Flight Research Center EC90-039-4 Photographed 1990
X-29 at an angle that highlights the forward swept wings.
NASA photo by Larry Sammons

The elastic axis is so far forward of the leading edge that the wings may be safely flown in a swept-forward orientation. This is in complete contrast to the earlier B-47 example that had swept-aft wings. Forward sweep is aerodynamically desirable because such wings have better stall characteristics, along with other advantages. However, this could not be safely realized until the use of carbon fiber construction.

4 Control Reversal

4.1 INTRODUCTION

The purpose of a control surface is to impart a force or moment about the aircraft's mass center, which initiates a motion to change the aircraft's orientation. The figure shows a deflected aileron on a wing. This produces a lift increment, ΔF_a, which, in turn, causes a rolling moment, ΔM_a, about the airplane's longitudinal axis:

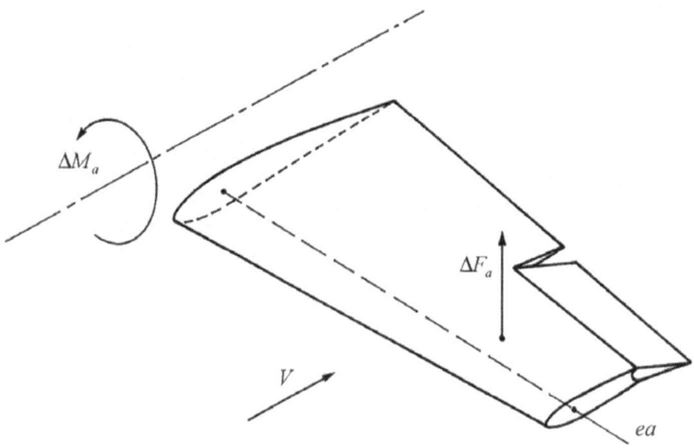

However, because the center of the lift increment, ΔF_a, is located behind the elastic axis, the wing can twist from this load and produce an opposing force and moment:

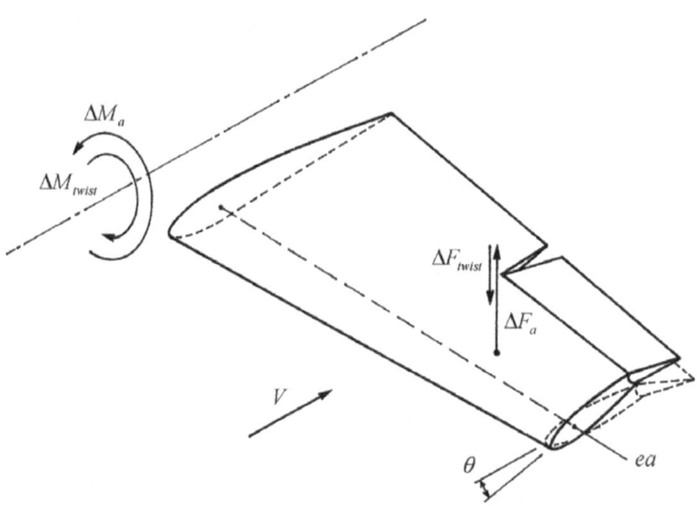

 DOI: 10.1201/9781032709093-4

This acts to reduce control effectiveness. If effectiveness is progressively reduced with increasing dynamic pressure, q, then "control reversal" is approached. When total ineffectiveness is reached, the corresponding dynamic pressure is called q_{rev}. Note that this is *not* an elastic stability problem as discussed for divergence.

4.2 TWO-DIMENSIONAL CASE

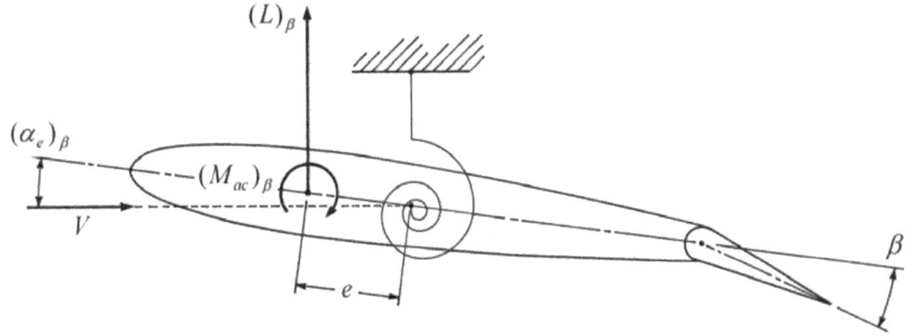

As discussed previously, the airfoil rotates from the aerodynamic load, L and M_{ac}, imposed upon it. In this case, there is a component of L and M_{ac} produced by the control deflection, β, that contributes to the elastic deflection, α_e. Therefore, the lift caused by the control deflection may be written as

$$(L)_\beta = qS(C_L)_\beta = qS\left[C_{L_\alpha}(\alpha_e)_\beta + C_{L_\beta}\beta\right] \tag{4.1}$$

where

$$C_{L_\alpha} = \partial C_L / \partial\alpha \ , \quad C_{L_\beta} = \partial C_L / \partial\beta$$

The torque about the elastic center due to the control deflection is

$$(T)_\beta \approxeq (L)_\beta\, e + (M_{ac})_\beta = qS\left\{e\left[C_{L_\alpha}(\alpha_e)_\beta + C_{L_\beta}\beta\right] + cC_{M_{ac}\beta}\beta\right\}$$

where

$$C_{M_{ac}\beta} = \partial\left(C_{M_{ac}}\right)/\partial\beta$$

which gives the elastic deflection

$$(\alpha_e)_\beta = C^{\theta\theta}(T)_\beta = C^{\theta\theta}qS\left\{e\left[C_{L_\alpha}(\alpha_e)_\beta + C_{L_\beta}\beta\right] + cC_{M_{ac}\beta}\beta\right\}$$

Upon solving for $(\alpha_e)_\beta$, one obtains

$$(\alpha_e)_\beta = \frac{C^{\theta\theta}qS\left[eC_{L_\beta} + cC_{M_{ac}\beta}\right]\beta}{1 - C^{\theta\theta}qSeC_{L_\alpha}}$$

Therefore, when this value of $(\alpha_e)_\beta$ is substituted into Equation (4.1), the lift component caused by the control deflection is

$$(L)_\beta = qSC_{L_\beta}\beta + qSC_{L_\alpha}\left\{\frac{C^{\theta\theta}qS\left[eC_{L_\beta}+cC_{M_{ac}\beta}\right]\beta}{1-C^{\theta\theta}qSeC_{L_\alpha}}\right\} =$$

$$(L)_\beta = \frac{qS\left[C_{L_\beta}+C^{\theta\theta}qScC_{L_\alpha}C_{M_{ac}\beta}\right]\beta}{1-C^{\theta\theta}qSeC_{L_\alpha}}$$

If the airfoil was perfectly rigid ($C^{\theta\theta}=0$), then the rigid lift would be given by

$$(L)_\beta^{rigid} = qSC_{L_\beta}\beta$$

Thus, the "control effectiveness" may be defined as

$$\frac{(L)_\beta}{(L)_\beta^{rigid}} = \frac{C_{L_\beta}+\left[C^{\theta\theta}qScC_{L_\alpha}C_{M_{ac}\beta}\right]}{C_{L_\beta}\left(1-C^{\theta\theta}qSeC_{L_\alpha}\right)} \tag{4.2}$$

Note that the denominator gives q_{div}, as before. Also, the control effectiveness goes to zero as $(L)_\beta \to 0$, so

$$q_{rev} = \frac{-1}{C^{\theta\theta}ScC_{L_\alpha}C_{M_{ac}\beta}/C_{L_\beta}} \tag{4.3}$$

One may also interpret control reversal in terms of the center-of-pressure movement due to control deflection. The following figure shows a symmetrical airfoil fixed at zero angle of attack (note that from linear airfoil theory, the aerodynamic solutions for camber and angle of attack are additive to this solution):

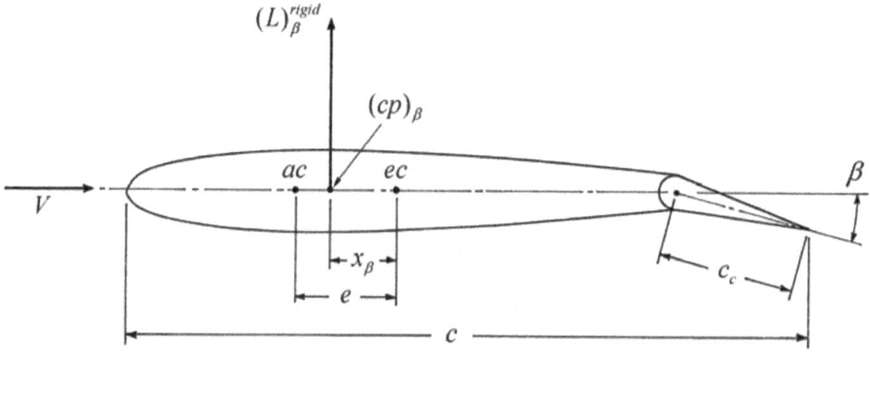

The pitching moment about the aerodynamic center is

$$\left(M_{ac}\right)_\beta = qScC_{M_{ac}\beta}\beta = \left(L\right)_\beta^{rigid}\left(x_\beta - e\right)$$

which gives that

$$\frac{C_{M_{ac}\beta}}{C_{L_\beta}} = \left[\frac{\left(x_\beta - e\right)\left(L\right)_\beta^{rigid}}{qSc}\right] / \left(\frac{\left(L\right)_\beta^{rigid}}{qS}\right) = \frac{\left(x_\beta - e\right)}{c}$$

Therefore, Equation (4.3) becomes

$$q_{rev} = \frac{1}{C^{\theta\theta}\, SC_{L\alpha}\left(e - x_\beta\right)} \tag{4.4}$$

From this, ones sees the following:

1. If $\left(cp\right)_\beta$ is at the ec, then $x_\beta = 0$ and $\left(L\right)_\beta^{rigid}$ produces no torque about the ec.

$$q_{rev} = q_{div}$$

2. If $\left(cp\right)_\beta$ is ahead of the ec, then $x_\beta \succ 0$ and $\left(L\right)_\beta^{rigid}$ produces a nose-up twist. One gets increased control effectiveness.

$$q_{rev} \succ q_{div}$$

3. If $\left(cp\right)_\beta$ is behind the ec, then $x_\beta \prec 0$ and $\left(L\right)_\beta^{rigid}$ produces a nose-down twist. Control effectiveness is reduced.

$$q_{rev} \prec q_{div}$$

The $\left(e - x_\beta\right)$ effect may be graphically summarized by using Equation (4.4) to write (4.2) as

$$\frac{\left(L\right)_\beta}{\left(L\right)_\beta^{rigid}} = \frac{1 - q/q_{rev}}{1 - q/q_{div}}$$

which plots as follows:

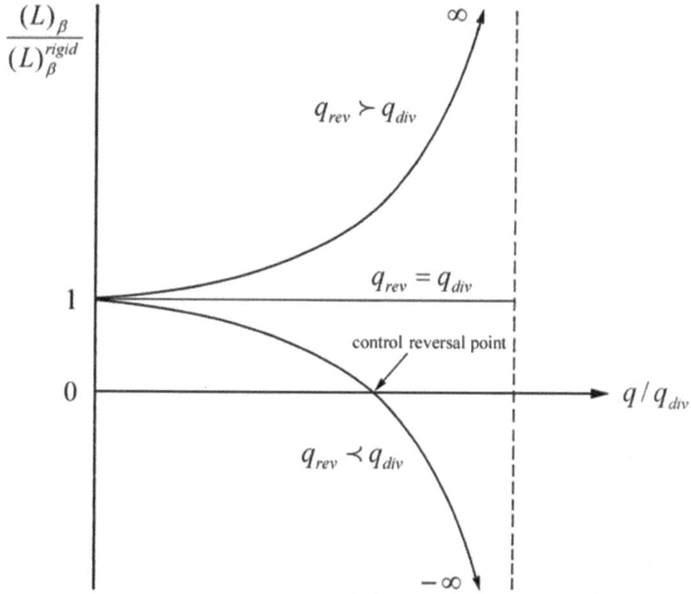

For a given elastic center location, the behavior of the section is determined by the center of pressure location. This is a function of the control size, c_c, relative to the chord, as well as the flow speed regime. Note that small c_c/c denotes a narrow "tab" at the trailing edge of the section, whereas $c_c/c = 1$ denotes an "all-moving section."

4.2.1 Subsonic

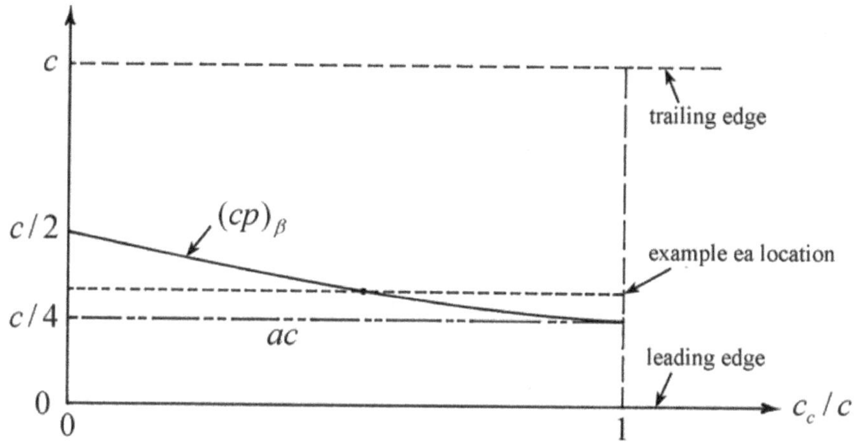

This shows that for the full range of c_c/c, $(cp)_\beta$ stays between the $c/2$ and $c/4$ locations. However, the elastic axis location is typically around the $c/3$ location. Therefore, one sees a crossover point where, for a large enough value of c_c/c, control reversal would become control augmentation (note that from an operational standpoint, neither situation is desirable).

4.2.2 Supersonic

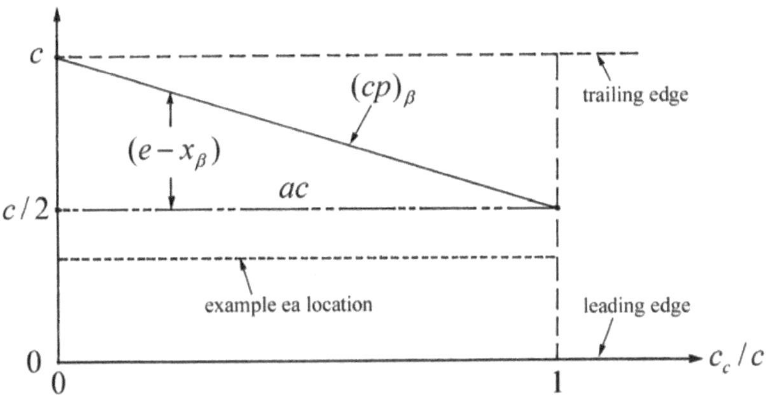

In this case, $(cp)_\beta$ varies between $c/2$ and c, which is almost always aft of the elastic axis. Therefore, control reversal is more of a problem for supersonic flight than subsonic. However, the lower value of C_{L_α} (which is about half that for subsonic flow) reduces the magnitude of the problem.

4.3 CANTILEVERED FINITE WING

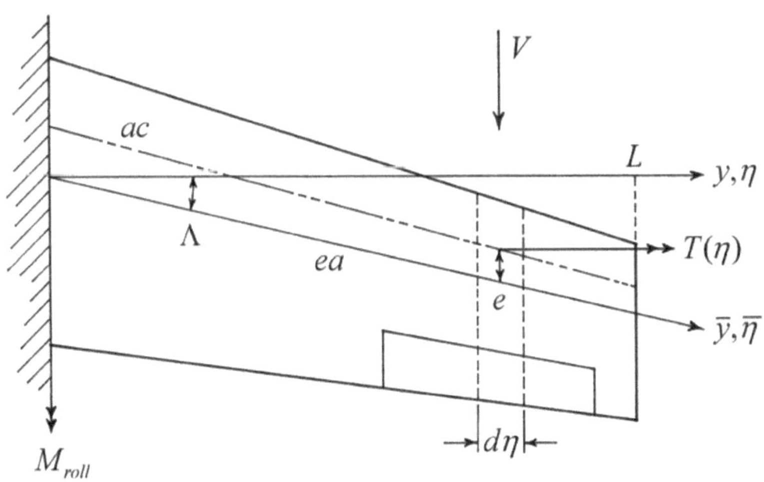

If one now considers C_{L_α}, C_{L_β}, $(\alpha_e)_\beta$, and β to be functions of y and, for simplicity, one defines $\alpha_e = (\alpha_e)_\beta$, then Equation (4.1) becomes

$$(L)_\beta = qc\left[C_{L_\alpha}\alpha_e + C_{L_\beta}\beta\right]$$

Further, e and $C_{Mac\beta}$ are likewise functions of y, so that the torsion equation becomes

$$(T)_{\beta} = qc\left\{e\left[C_{L_\alpha}\alpha_e + C_{L_\beta}\beta\right] + cC_{M_{ac}\beta}\beta\right\}$$

From Equation (2.5), the elastic deformation may be expressed as

$$\alpha_e = \int_0^L C^{\theta\theta}(y,\eta)(T)_{\beta}\,d\eta + \int_0^L C^{\theta z}(y,\eta)(L)_{\beta}\,d\eta$$

These equations combine to give

$$\alpha_e = q\int_0^L C_{L_\alpha} c\left[C^{\theta\theta}(y,\eta)e + C^{\theta z}(y,\eta)\right]\alpha_e\,d\eta$$

$$+\,q\int_0^L C_{L_\beta} c\left[C^{\theta\theta}(y,\eta)e + C^{\theta z}(y,\eta)\right]\beta\,d\eta + q\int_0^L C^{\theta\theta}(y,\eta)c^2 C_{M_{ac}\beta}\beta\,d\eta$$

This may be expressed in a discrete-element formulation as

$$(\alpha_e)_i = q\sum_{j=1}^{n}\tilde{A}_{ij}(\alpha_e)_j + q\sum_{j=1}^{n}\tilde{B}_{ij}\beta_j$$

where

$$\tilde{A}_{ij} = \left(C_{L_\alpha}\right)_j\left(C_{ij}^{\theta\theta}e_j + C_{ij}^{\theta z}\right)c_j\,\delta\eta_j \tag{4.5}$$

$$\tilde{B}_{ij} = C_{ij}^{\theta\theta}c_j^2\left(C_{Mac\beta}\right)_j\delta\eta_j + \left(C_{L_\beta}\right)_j\left(C_{ij}^{\theta\theta}e_j + C_{ij}^{\theta z}\right)c_j\,\delta\eta_j \tag{4.6}$$

In matrix form, this becomes

$$\{\alpha_e\} = q\left[\tilde{A}\right]\{\alpha_e\} + q\left[\tilde{B}\right]\{\beta\}$$

$$\left[\tilde{A}\right] = \left[C^{\theta\theta}\right][e]\left[C_{L_\alpha}\right][c][\delta\eta] + \left[C^{\theta z}\right]\left[C_{L_\alpha}\right][c][\delta\eta] \tag{4.7}$$

$$\left[\tilde{B}\right] = \left[C^{\theta\theta}\right][c][c]\left[C_{M_{ac}\beta}\right][\delta\eta] + \left[C^{\theta\theta}\right][e]\left[C_{L_\beta}\right][c][\delta\eta] + \left[C^{\theta z}\right]\left[C_{L_\beta}\right][c][\delta\eta] \tag{4.8}$$

which may be rearranged to give a solution for $\{\alpha_e\}$:

$$\{\alpha_e\} = q\left[I - q\tilde{A}\right]^{-1}\left[\tilde{B}\right]\{\beta\}$$

Note that the matrices $[e]$, $\left[C_{L_\alpha}\right]$, $[c]$, $[\delta\eta]$, $\left[C_{M_{ac}\beta}\right]$, and $\left[C_{L_\beta}\right]$ are diagonal. For example:

$$[e] = \begin{bmatrix} e_1 & 0 & 0 & 0 \\ 0 & e_2 & 0 & 0 \\ . & . & . & . \\ 0 & 0 & . & e_n \end{bmatrix}, \quad \left[C_{L_\alpha}\right] = \begin{bmatrix} \left(C_{L_\alpha}\right)_1 & 0 & 0 & 0 \\ 0 & \left(C_{L_\alpha}\right)_2 & 0 & 0 \\ . & . & . & . \\ 0 & 0 & . & \left(C_{L_\alpha}\right)_n \end{bmatrix}, \text{ etc.}$$

Now, the total rolling moment of both wing halves about the fuselage axis is given by

$$M_{roll} = 2\int_0^L y(L)_\beta \, dy$$

which, in discrete-element form, is

$$M_{roll} = 2\sum_{i=1}^{n} y_i (L_\beta)_i \, \delta y_i$$

where

$$(L_\beta)_i = q c_i \left[(C_{L_\alpha})_i (\alpha_e)_i + (C_{L_\beta})_i \beta_i \right]$$

This may be expressed in matrix form:

$$M_{roll} = 2q \times sum\left([y][c][C_{L_\alpha}][\delta y]\{\alpha_e\}\right) + 2q \times sum\left([y][c][C_{L_\beta}][\delta y]\{\beta\}\right)$$

where $[y]$ and $[\delta y]$ are diagonal matrices. From this, the rolling moment for a rigid (non-elastic) wing is obtained:

$$(M_{roll})_{rigid} = 2q \times sum\left([y][c][C_{L_\beta}][\delta y]\{\beta\}\right)$$

And one may define the control effectiveness as

$$\text{control effectiveness} = M_{roll} / (M_{roll})_{rigid}$$

4.3.1 Numerical Example

Consider the uniform cantilevered wing shown here:

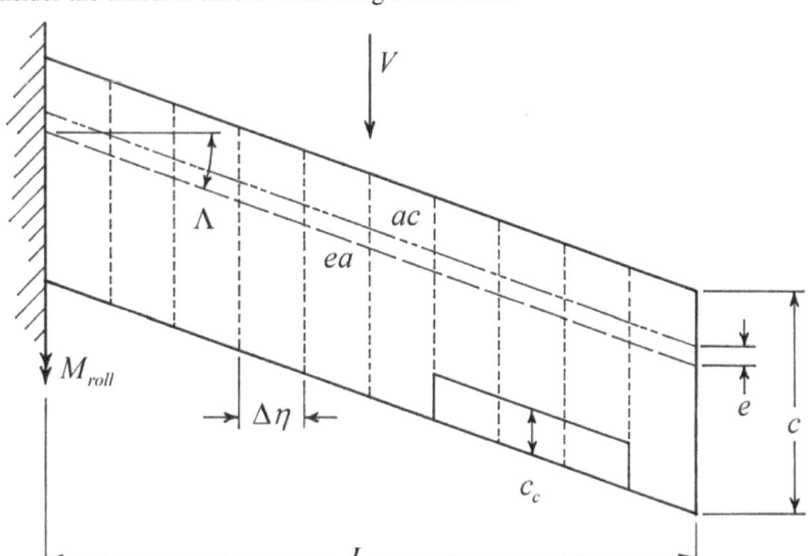

This is the same wing as in Example 3.2, only now with a control surface. As before,

$$e = \bar{e} / \cos\Lambda,\ c = \bar{c} / \cos\Lambda$$

where $\bar{e} = 0.1524m$ and $\bar{c} = 1.8288m$. Also, using the notation of this section,

$$C_{L_\alpha} = \bar{C}_{L_\alpha} \cos\Lambda,\ C_{L_\beta} = \bar{C}_{L_\beta} \cos\Lambda,\ C_{Mac\beta} = \bar{C}_{Mac\beta} \cos\Lambda$$

where $\bar{C}_{L_\alpha} = 4.775$ and thin airfoil theory gives, for a simple flapped section, that

$$\bar{C}_{L_\beta} = 2\left[\left(\pi - \tilde{\theta}\right) + \sin\tilde{\theta}\right],\ \bar{C}_{Mac\beta} = \left[\sin\tilde{\theta}\left(\cos\tilde{\theta} - 1\right)\right] / 2,\ \tilde{\theta} = \cos^{-1}\left[1 - 2\left(1 - c_c / c\right)\right]$$

For the 20% ailerons of this example, $\bar{C}_{L_\beta} = 3.4546$ and $\bar{C}_{Mac\beta} = -0.640$.

Further, the span is given by $L = \bar{L}\cos\Lambda$, where $\bar{L}$ is the length of the elastic axis, $\bar{L} = 6.096m$. Also, the $\delta\eta$ increments are

$$\delta\eta = \delta\bar{\eta}\cos\Lambda,\ \delta\bar{\eta} = 0.6096m$$

Finally, the uniform elastic properties along the elastic axis are

$$EI\left(\bar{y}\right) = 9.7523 \times 10^6\ N - m^2;\ GJ\left(\bar{y}\right) = 9.8763 \times 10^5\ N - m^2 / rad$$

from which the flexibility influence-coefficient values may be calculated as described in Chapter 2.

For the following conditions,

$$\rho = 1.225kg / m^3,\ \Lambda = 20 \text{ degrees},\ V = 100m / \sec$$
$$\beta_1 \text{ through } \beta_6 = 0,\ \beta_7,\ \beta_8,\ \beta_9 = 0.01rad,\ \beta_{10} = 0$$

one obtains the elastic twist:

$$\{\alpha_e\} = \begin{Bmatrix} -0.0394 \\ -0.1152 \\ -0.1869 \\ -0.2543 \\ -0.3172 \\ -0.3751 \\ -0.4279 \\ -0.4621 \\ -0.4791 \\ -0.4803 \end{Bmatrix} (\text{deg})$$

which gives a total rolling moment of

$$M_{roll} = 44935N - m$$

By comparison, the rigid-wing rolling moment is

$$(M_{roll})_{rigid} = 57140N - m$$

Therefore,

$$\text{control effectiveness} = 0.7864$$

It is clear that, at this speed, control reversal is not a problem for this wing. In fact, the speed has to increase to $200m/\sec$ for control effectiveness to reach zero. This is beyond the validity of the incompressible-flow aerodynamics assumed for this example. However, if $EI(y)$ is reduced to 10% of its original value, then $V_{reversal}$ becomes $179m/\sec$. For thin wings that are susceptible to reversal, spoilers are often used instead of ailerons because $C_{Mac\beta}$ is much less for a given value of C_{L_β}. Spoilers were tested on the Boeing B-47 bomber described in Section 3.3.1 and were subsequently incorporated into the B-52 bomber and modern jet airliners.

4.4 DYNAMIC CONTROL REVERSAL

Up to this point, only static control reversal has been considered. However, the consequence of a control input is the dynamic response of the aircraft. The following figure shows a wing undergoing a roll rate, P, (angular velocity) about an aircraft fuselage's centerline. For simplicity, the fuselage is not shown because, assuming a slender shape, its location near $y \approx 0$ gives it a negligible effect on the roll rate results.

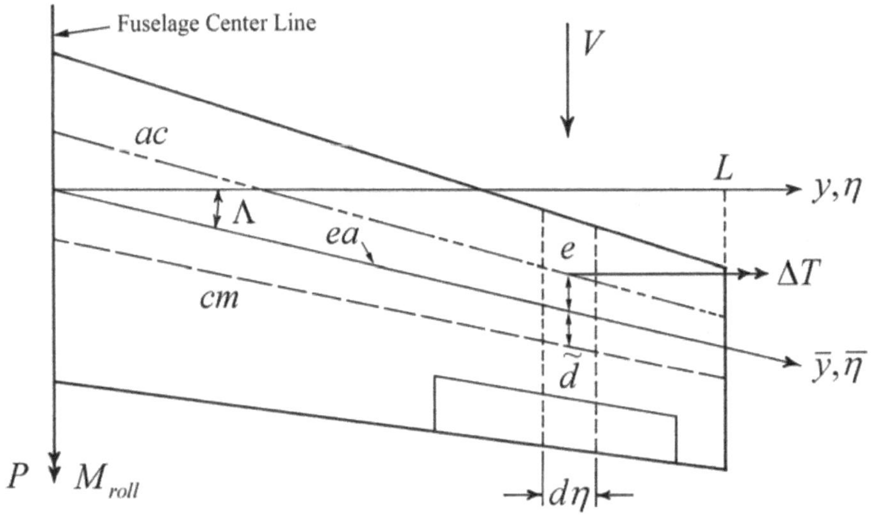

Although one could calculate the time history of the roll, $P(t)$, there are two special cases that characterize the roll behavior of this aircraft: the steady-state roll rate and the instantaneous roll acceleration due to sudden control input.

4.4.1 Steady-State Roll Rate

For this case, the question is, after the initial transient response, what roll rate does the aircraft converge to? The first step in finding this is to note that each section along the wing experiences an additional aerodynamic angle of attack, α_{roll}, due to the roll rate:

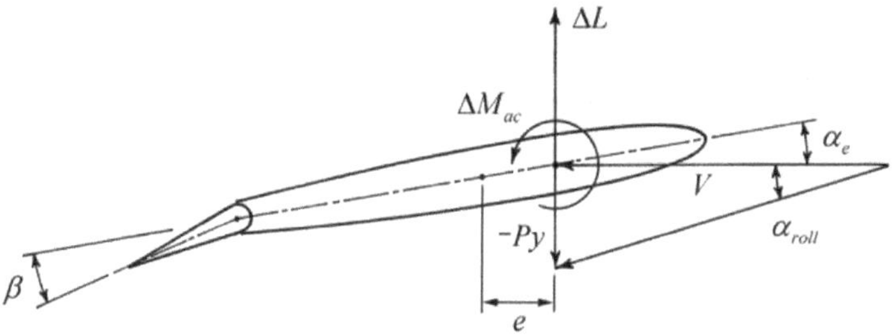

$$\alpha_{roll} = -Py / V$$

Therefore, the lift force and torsional moment along the wing are given by

$$\Delta L(y) = qc\left[C_{L_\alpha}\alpha_e + C_{L_\beta}\beta - C_{L_\alpha}(Py/V)\right]$$

$$\Delta T(y) = e\Delta L + qc^2 C_{M_{ac\beta}}\beta = qce\left[C_{L_\alpha}\alpha_e + C_{L_\beta}\beta - C_{L_\alpha}(Py/V)\right] + qc^2 C_{M_{ac\beta}}\beta$$

As before, the elastic deformation is obtained from

$$\alpha_e = \int_0^L C^{\theta\theta}(y,\eta)\Delta T(\eta)d\eta + \int_0^L C^{\theta z}(y,\eta)\Delta L(\eta)\, d\eta$$

so that these equations combine to give

$$\begin{aligned}
\alpha_e = {} & q\int_0^L C_{L_\alpha} c\left[C^{\theta\theta}(y,\eta)e + C^{\theta z}(y,\eta)\right]\alpha_e\, d\eta \\
& + q\int_0^L C_{L_\beta} c\left[C^{\theta\theta}(y,\eta)e + C^{\theta z}(y,\eta)\right]\beta\, d\eta \\
& + q\int_0^L C^{\theta\theta}(y,\eta)c^2 C_{M_{ac\beta}}\beta\, d\eta \\
& - q\int_0^L C_{L_\alpha} c\left[C^{\theta\theta}(y,\eta)e + C^{\theta z}(y,\eta)\right](P\eta/V)d\eta
\end{aligned} \tag{4.9}$$

Note that one may write

$$\frac{P\eta}{V} = \frac{\eta}{L}\left(\frac{PL}{V}\right)$$

so that, in discrete-element form, Equation (4.9) becomes

$$(\alpha_e)_i = q\sum_{j=1}^{n}\tilde{A}_{ij}(\alpha_e)_j + q\sum_{j=1}^{n}\tilde{B}_{ij}\beta_j - q\sum_{j=1}^{n}\tilde{C}_{ij}\frac{\eta_j}{L}\left(\frac{PL}{V}\right)$$

where $\tilde{A}_{ij}$ and $\tilde{B}_{ij}$ are obtained from Equations (4.5) and (4.6), and

$$\tilde{C}_{ij} = \tilde{A}_{ij}$$

These equations may also be expressed in matrix form:

$$\{\alpha_e\} = q[\tilde{A}]\{\alpha_e\} + q[\tilde{B}]\{\beta\} - \frac{q}{L}[\tilde{C}]\{y\}\left(\frac{PL}{V}\right) \tag{4.10}$$

where $[\tilde{A}]$ and $[\tilde{B}]$ are obtained from Equations (4.7) and (4.8), and

$$[\tilde{C}] = [\tilde{A}]$$

Now, from before,

$$M_{roll} = 2\int_0^L y\Delta L(y)dy \approx 2\sum_{i=1}^{n} y_i(\Delta L)_i \delta y_i \tag{4.11}$$

where

$$(\Delta L)_i = qc_i\left[(C_{L_\alpha})_i(\alpha_e)_i + (C_{L_\beta})_i\beta_i - (C_{L_\alpha})_i\frac{y_i}{L}\left(\frac{PL}{V}\right)\right]$$

So in matrix form, this into Equation (4.11) gives

$$\begin{aligned} M_{roll} &= 2q \times sum\left([y][c][C_{L_\alpha}][\delta y]\{\alpha_e\}\right) + 2q \times sum\left([y][c][C_{L_\beta}][\delta y]\{\beta\}\right) \\ &\quad - \frac{2q}{L} \times sum\left([y][c][C_{L_\alpha}][\delta y]\{y\}\right)\left(\frac{PL}{V}\right) \end{aligned} \tag{4.12}$$

where $[y]$ and $[\delta y]$ are diagonal. Next, Equation (4.10) may be solved for α_e:

$$\{\alpha_e\} = q[I - q\tilde{A}]^{-1}[\tilde{B}]\{\beta\} - q\frac{1}{L}\left(\frac{PL}{V}\right)[I - q\tilde{A}]^{-1}[\tilde{C}]\{y\}$$

And this may be substituted into Equation (4.12) to give

$$\begin{aligned} M_{roll} &= 2q \times sum\left([\tilde{D}]\{\beta\} - \frac{1}{L}\left(\frac{PL}{V}\right)\{\tilde{E}\}\right) + 2q \times sum\left([y][c][C_{L_\beta}][\delta y]\{\beta\}\right) \\ &\quad - 2\frac{q}{L} \times sum\left([y][c][C_{L_\alpha}][\delta y]\{y\}\right)\left(\frac{PL}{V}\right) \end{aligned} \tag{4.13}$$

where

$$[\tilde{D}] = q[y][c][C_{L_\alpha}][\delta y][I - q\tilde{A}]^{-1}[\tilde{B}]$$

$$\{\tilde{E}\} = q[y][c][C_{L_\alpha}][\delta y][I - q\tilde{A}]^{-1}[\tilde{C}]\{y\}$$

These terms combine in Equation (4.13) to give

$$M_{roll} = 2q \times sum\left(\left[\tilde{F}\right]\{\beta\}\right) - 2\frac{q}{L}\left(\frac{PL}{V}\right) \times sum\left(\left\{\tilde{G}\right\}\right) \tag{4.14}$$

where

$$\left[\tilde{F}\right] = \left[\left[\tilde{D}\right] + \lfloor y \rfloor \lfloor c \rfloor \left[C_{L_\beta}\right] \lfloor \delta y \rfloor\right]$$

$$\left\{\tilde{G}\right\} = \left\{\tilde{E}\right\} + \lfloor y \rfloor \lfloor c \rfloor \left[C_{L_\alpha}\right] \lfloor \delta y \rfloor \{y\}$$

Now, for steady-state roll, one has that

$$M_{roll} = 0$$

so that Equation (4.14) becomes

$$\frac{1}{L}\left(\frac{PL}{V}\right) \times sum\left(\left\{\tilde{G}\right\}\right) = sum\left(\left[\tilde{F}\right]\{\beta\}\right) \tag{4.15}$$

Further, represent $\{\beta\}$ as

$$\{\beta\} = \{1_a\}\beta$$

where $\{1_a\}$ is a matrix representation of the Heaviside function. For example:

$$\{1_a\} = \begin{Bmatrix} 0 \\ 0 \\ . \\ . \\ 0 \\ 1 \\ 1 \\ 0 \end{Bmatrix}$$

where the *zero* values correspond to those wing segments with no control surfaces, and the *unity* values are for those segments that include a control surface. So Equation (4.15) may be written as

$$\frac{1}{L}\left(\frac{PL}{V}\right) \times sum\left(\left\{\tilde{G}\right\}\right) = sum\left(\left[\tilde{F}\right]\{1_a\}\right)\beta$$

and rearranged to give

$$\left(\frac{PL}{V}\right) = L\frac{sum\left(\left[\tilde{F}\right]\{1_a\}\right)\beta}{sum\left(\left\{\tilde{G}\right\}\right)}$$

The roll rate parameter may then be defined as

$$\frac{\partial(PL/V)}{\partial\beta}=L\frac{sum\left(\left[\tilde{F}\right]\{1_a\}\right)}{sum\left(\{\tilde{G}\}\right)} \tag{4.16}$$

For the rigid (non-twisting) wing, one has that

$$\Delta L_{rigid}=q\,c\left[C_{L_\beta}\,\beta-C_{L_\alpha}\,(Py/V)\right]\rightarrow$$

$$\left(\Delta L_{rigid}\right)_i=q\,c_i\left[\left(C_{L_\beta}\right)_i\beta_i-\left(C_{L_\alpha}\right)_i\frac{y_i}{L}\left(\frac{PL}{V}\right)\right]$$

Upon substitution into Equation (4.11), one obtains

$$\begin{aligned}\left(M_{roll}\right)_{rigid}&=2q\times sum\left([y][c]\left[C_{L_\beta}\right][\delta y]\{\beta\}\right)\\&-\frac{2q}{L}\times sum\left([y][c]\left[C_{L_\alpha}\right][\delta y]\{y\}\right)\left(\frac{PL}{V}\right)\end{aligned} \tag{4.17}$$

Define:

$$\left[\tilde{H}\right]=[y][c]\left[C_{L_\beta}\right][\delta y],\quad \{\tilde{J}\}=[y][c]\left[C_{L_\alpha}\right][\delta y]\{y\}$$

Also, for steady-state roll,

$$\left(M_{roll}\right)_{rigid}=0$$

so that Equation (4.17) gives

$$sum\left(\left[\tilde{H}\right]\{\beta\}\right)=sum\left(\{\tilde{J}\}\right)\frac{1}{L}\left(\frac{PL}{V}\right)\rightarrow\left(\frac{PL}{V}\right)_{rigid}=L\frac{sum\left(\left[\tilde{H}\right]\{1_a\}\right)\beta}{sum\left(\{\tilde{J}\}\right)}$$

from which the rigid-wing roll rate parameter is

$$\frac{\partial(PL/V)_{rigid}}{\partial\beta}=L\frac{sum\left(\left[\tilde{H}\right]\{1_a\}\right)}{sum\left(\{\tilde{J}\}\right)} \tag{4.18}$$

For the previous numerical example, Equation (4.18) gives the steady-state rigid-wing roll rate at $100m/\sec$ as

$$\frac{\partial(PL/V)_{rigid}}{\partial\beta}=0.4896rad/\sec$$

Further, the flexible-wing roll rates were calculated for a range of speeds and are shown on the figure as a ratio of the rigid-wing values:

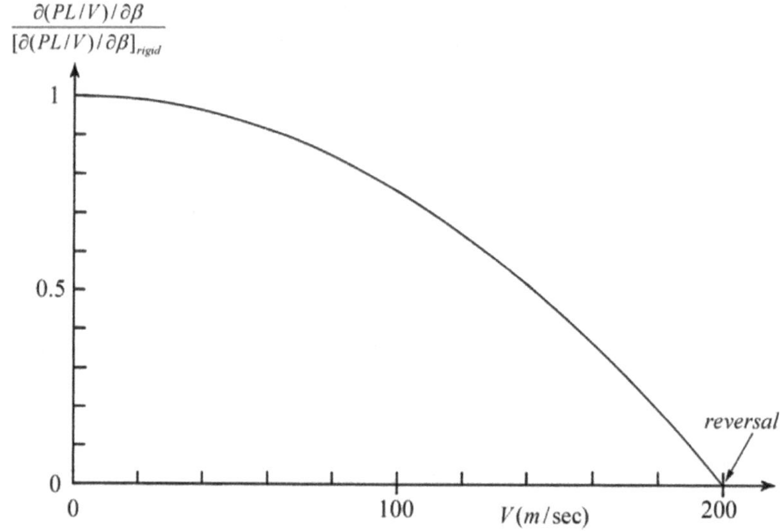

It is seen that the speed for reversal, $200 m/\sec$, is exactly the same as that for the previous example. This makes sense because, by definition, this is the case where the roll rate is zero and α_{roll} is zero along the span. Hence, the situation is exactly the same as that for the static cantilevered case. Finally, note that this curve may be closely approximated by

$$\frac{\partial(PL/V)/\partial\beta}{\left[\partial(PL/V)/\partial\beta\right]_{rigid}} \approx 1 - \frac{q}{q_{rev}} = 1 - \left(\frac{V}{V_{rev}}\right)^2$$

4.4.2 Initial Roll Acceleration

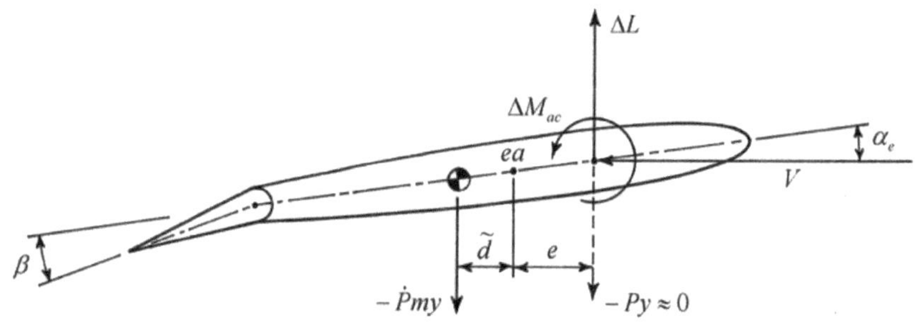

In this case, the question is, if the controls are suddenly applied, what is the initial roll acceleration, $\dot{P}$? Note that $P \approx 0$ in this first instant so that the values of the vertical force, $\Delta Z(y)$, and the twisting moment, $\Delta T(y)$, are

$$\Delta Z(y) = \Delta L - \dot{P}my = qc\left[C_{L_\alpha}\alpha_e + C_{L_\beta}\beta\right] - \dot{P}my$$

$$\Delta T(y) = e\Delta L + qc^2 C_{M_{ac\beta}}\beta + \dot{P}my\tilde{d} = qce\left[C_{L_\alpha}\alpha_e + C_{L_\beta}\beta\right] + qc^2 C_{M_{ac\beta}}\beta + \dot{P}my\tilde{d}$$

As before, the elastic deformation is given by

$$\alpha_e = \int_0^L C^{\theta\theta}(y,\eta)\Delta T(\eta)d\eta + \int_0^L C^{\theta z}(y,\eta)\Delta Z(\eta)\,d\eta$$

and these equations combine to give

$$\begin{aligned}\alpha_e &= q\int_0^L C_{L_\alpha} c\left[C^{\theta\theta}(y,\eta)e + C^{\theta z}(y,\eta)\right]\alpha_e\,d\eta \\ &+ q\int_0^L C_{L_\beta} c\left[C^{\theta\theta}(y,\eta)e + C^{\theta z}(y,\eta)\right]\beta\,d\eta \\ &+ q\int_0^L C^{\theta\theta}(y,\eta)c^2 C_{M_{ac\beta}}\beta\,d\eta \\ &+ \dot{P}\int_0^L m\left[C^{\theta\theta}(y,\eta)\tilde{d} - C^{\theta z}(y,\eta)\right]\eta\,d\eta\end{aligned} \tag{4.19}$$

In discrete element form, Equation (4.19) becomes

$$(\alpha_e)_i = q\sum_{j=1}^{n}\tilde{A}_{ij}(\alpha_e)_j + q\sum_{j=1}^{n}\tilde{B}_{ij}\beta_j + \dot{P}\sum_{j=1}^{n}\hat{C}_{ij}\eta_j$$

where $\tilde{A}_{ij}$ and $\tilde{B}_{ij}$ are obtained from Equations (4.5) and (4.6), and

$$\hat{C}_{ij} = m_j\left(C_{ij}^{\theta\theta}\tilde{d}_j - C_{ij}^{\theta z}\right)$$

These equations may also be written in matrix form:

$$\{\alpha_e\} = q\left[\tilde{A}\right]\{\alpha_e\} + q\left[\tilde{B}\right]\{\beta\} + \dot{P}\left[\hat{C}\right]\{y\} \tag{4.20}$$

where $\left[\tilde{A}\right]$ and $\left[\tilde{B}\right]$ are obtained from Equations (4.7) and (4.8), and

$$\left[\hat{C}\right] = \left[C^{\theta\theta}\right]\left[\tilde{d}\right][m] - \left[C^{\theta z}\right][m]$$

Note that $[m]$ and $\left[\tilde{d}\right]$ are diagonal, as well as the previously defined diagonal matrices.

Now, from before, the imposed rolling moment from the aerodynamic forces is

$$M_{roll} = 2\int_0^L y\Delta L(y)dy \approx 2\sum_{i=1}^{n} y_i(\Delta L)_i\,\delta y_i \tag{4.21}$$

where

$$(\Delta L)_i = qc_i\left[\left(C_{L_\alpha}\right)_i(\alpha_e)_i + \left(C_{L_\beta}\right)_i\beta_i\right]$$

This and Equation (4.21) gives, in matrix form,

$$M_{roll} = 2q\times sum\left([y][c]\left[C_{L_\alpha}\right][\delta y]\{\alpha_e\}\right) + 2q\times sum\left([y][c]\left[C_{L_\beta}\right][\delta y]\{\beta\}\right) \tag{4.22}$$

Next, Equation (4.20) is solved for $\{\alpha_e\}$:

$$\{\alpha_e\} = q\left[I - q\tilde{A}\right]^{-1}\left[\tilde{B}\right]\{\beta\} + \dot{P}\left[I - q\tilde{A}\right]^{-1}\left[\hat{C}\right]\{y\}$$

And this may be substituted into Equation (4.22):

$$M_{roll} = 2q\times sum\left(\left[\tilde{D}\right]\{\beta\} + \dot{P}\left\{\hat{E}\right\}\right) + 2q\times sum\left([y][c]\left[C_{L_\beta}\right][\delta y]\{\beta\}\right) \tag{4.23}$$

where

$$\left[\tilde{D}\right] = q[y][c]\left[C_{L_\alpha}\right][\delta y]\left[I - q\tilde{A}\right]^{-1}\left[\tilde{B}\right]$$

$$\left\{\hat{E}\right\} = [y][c]\left[C_{L_\alpha}\right][\delta y]\left[I - q\tilde{A}\right]^{-1}\left[\hat{C}\right]\{y\}$$

The terms combine in Equation (4.23) to give

$$M_{roll} = 2q\times sum\left(\left[\tilde{F}\right]\{\beta\}\right) + 2q\dot{P}\times sum\left(\left\{\hat{E}\right\}\right) \tag{4.24}$$

where

$$\left[\tilde{F}\right] = \left[\left[\tilde{D}\right] + [y][c]\left[C_{L_\beta}\right][\delta y]\right]$$

Now, the initial roll acceleration is given by

$$\left(\dot{P}\right)_{initial} = \frac{M_{roll}}{I_{xx}} \tag{4.25}$$

where I_{xx} is the total rolling moment of inertia of the airplane. Equation (4.24) may be substituted into (4.25) to give

$$\left(\dot{P}\right)_{initial} = \frac{2q}{I_{xx}}\times sum\left(\left[\tilde{F}\right]\{\beta\}\right) + \frac{2q\dot{P}}{I_{xx}}\times sum\left(\left\{\hat{E}\right\}\right) \tag{4.26}$$

Upon writing $\{\beta\}$ as $\{1_a\}\beta$, where $\{1_a\}$ is previously defined, one may solve for $\left(\dot{P}\right)_{initial}$:

$$\left(\dot{P}\right)_{initial} = \frac{2q \times sum\left(\left[\tilde{F}\right]\{1_a\}\right)\beta}{I_{xx} - 2q \times sum\left(\left\{\hat{E}\right\}\right)} \tag{4.27}$$

4.4.3 Numerical Example

Consider that the uniform swept wing described in the previous numerical example has a uniform mass distribution whose center is located at

$$\tilde{d} = \bar{d} / \cos\Lambda \text{ where } \bar{d} = 0.1826m$$

Further, the discrete mass values, m_i, for each segment are given by

$$m_i = m(\bar{\eta})\delta\bar{\eta} = m(\bar{\eta})\delta\eta / \cos\Lambda$$

where

$$m(\bar{\eta}) = 35.7087 kg / m$$

For the following parameters,

$$\Lambda = 20 \text{ degrees}, V = 100 m / \sec, \beta = 0.1 \text{ rad}$$

the elastic twist distribution is found to be

$$\{\alpha_e\} = \begin{Bmatrix} -0.0084 \\ -0.0252 \\ -0.0425 \\ -0.0605 \\ -0.0798 \\ -0.1006 \\ -0.1230 \\ -0.1341 \\ -0.1355 \\ -0.1290 \end{Bmatrix} (\text{deg})$$

Compared with the static control reversal case, the $\{\alpha_e\}$ distribution is seen to be less negative. This is a consequence of the aft mass center location causing the acceleration to produce a positive twisting moment at each section. This is seen in the resulting rolling moment:

$$\left(M_{roll}\right)_{initial} = 53798N - m$$

which is 20% higher than the static value.

Finally, for a total rolling moment of inertia, $I_{xx} = 6000$ kg-m², one obtains

$$\dot{P} = 8.9663 rad / \sec^2$$

5 Structural Dynamics

5.1 INTRODUCTION

The purpose of this chapter is to describe various methods for obtaining the resonant vibration modes, in torsion and bending, of a straight structural element. Such information will prove useful for the flutter analysis of finite wings (Section 6.4), where the assumed mode solutions are based on the unforced undamped modes described in the following sections.

5.2 TORSIONAL VIBRATION

A segment of rod is shown as follows, upon which is imposed a distributed torque $t(y)$:

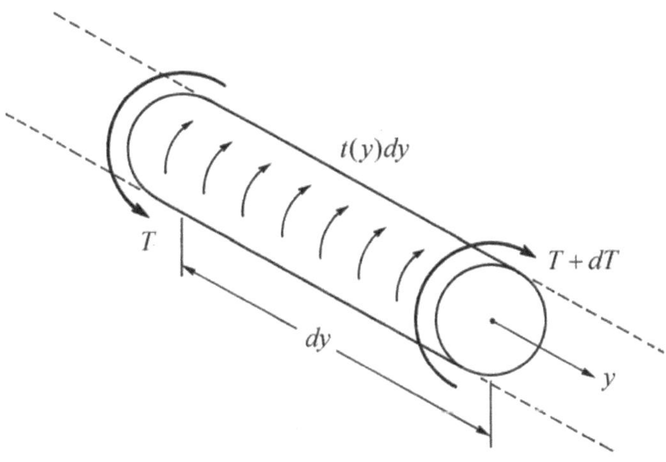

The equilibrium equation is

$$T + dT + t(y)dy = T \rightarrow dT + t(y)dy = 0 \rightarrow \frac{dT}{dy} = -t(y) \tag{5.1}$$

Further, the torsional elastic equation is

$$T = GJ\frac{d\theta}{dy} \tag{5.2}$$

So Equations (5.1) and (5.2) combine to give

$$-t(y) = \frac{d}{dy}\left(GJ\frac{d\theta}{dy}\right) \tag{5.3}$$

DOI: 10.1201/9781032709093-5

Now, let t be a function of time as well, $t(y,t)$. If no external forces act except for rotational inertia, and the elastic axis of the rod coincides with its cross-sectional mass center, then $t(y,t)$ is

$$t(y,t) = -i_{cm}\,\ddot{\theta}$$

where $i_{cm} \equiv$ rotational mass moment of inertia per unit length about the y axis, and $\ddot{\theta}$ is the second partial derivative of θ with respect to time. Equation (5.3) thus becomes

$$\frac{\partial}{\partial y}\left(GJ\frac{\partial \theta}{\partial y}\right) = i_{cm}\,\ddot{\theta} \tag{5.4}$$

As a special case, choose $GJ = Const.$ and $i_{cm}(y) = Const.$ so that Equation (5.4) takes the form

$$GJ\frac{\partial^2 \theta}{\partial y^2} = i_{cm}\,\ddot{\theta} \tag{5.5}$$

Assume a separation of variables for the solution

$$\theta(y,t) = \Theta(y)\,func(t) = \Theta(y)\sin\omega t$$

This, into Equation (5.5), gives

$$\frac{d^2\Theta}{dy^2} = -\frac{i_{cm}\,\omega^2}{GJ}\Theta \equiv -k^2\Theta$$

for which the general solution is

$$\Theta(y) = A\sin ky + B\cos ky \tag{5.6}$$

5.2.1 Solution for a Uniform Cantilever Rod

The end conditions for the cantilever rod are

$$\theta(0) = 0 \rightarrow \Theta(0) = 0 \rightarrow B = 0 \text{ and } \theta'(L) = 0 \rightarrow (d\Theta/dy)_L = 0 \rightarrow A\cos(kL) = 0$$

The second end condition gives that either $A = 0$ or, for a non-trivial solution, $\cos(kL) = 0$. Thus,

$$k_j = \frac{a_j}{L} \text{ where } a_j = (2j-1)\frac{\pi}{2}$$

From the preceding definition for k, the natural (unforced) frequencies are then obtained:

$$\omega_j = \frac{a_j}{L}\left(\frac{GJ}{i_{cm}}\right)^{1/2}$$

Next, Equation (5.6) gives the mode shapes corresponding to these frequencies:

$$\Theta_j(y) = A_j \sin\left(\frac{a_j}{L}y\right)$$

The first three modes are drawn as follows. Note that these are normalized with respect to A_j so that the maximum magnitudes are unity. Also, observe that these satisfy the end conditions of zero displacement at $y = 0$ and zero slope at L.

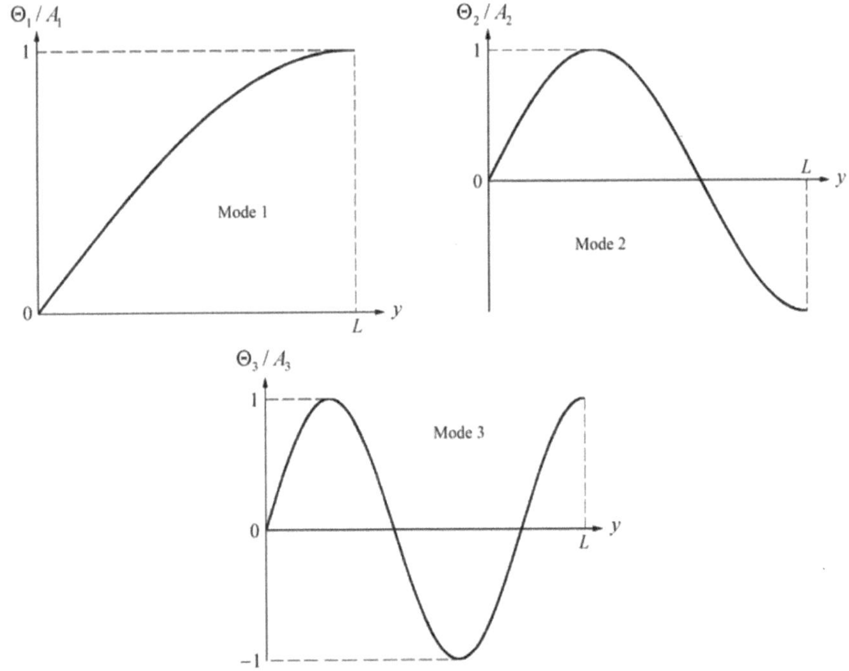

Although the first mode shape is the same as that for the previously discussed uniform wing undergoing divergence, generally the dynamic mode shapes are different than the static divergence shapes. Also, a feature of the dynamics of continuous structures is an infinity of distinct frequencies and modes (transcendental characteristic equation).

5.2.2 Solution for a Uniform Free-Free Rod

The end conditions for a free-free rod are

$$\theta'(0) = 0, \theta'(L) = 0$$

from which Equation (5.6) gives that

$$(d\Theta / dy)_0 = 0 \rightarrow A = 0 \text{ and } (d\Theta / dy)_L = 0 \rightarrow B\sin(kL) = 0$$

For a non-trivial solution, one has that

$$\sin(kL) = 0 \rightarrow k_j = \frac{a_j}{L} \text{ where } a_j = \pi j,\ j = 1, 2, \ldots N$$

From the previous definition for k, the unforced vibratory frequencies are

$$\omega_j = \frac{a_j}{L}\left(\frac{GJ}{i_{cm}}\right)^{1/2}$$

Further, Equation (5.6) gives the mode shapes corresponding to these frequencies:

$$\Theta_j(y) = B_j \cos\left(\frac{a_j}{L} y\right)$$

The first three vibratory modes are shown as follows, with the origin of the y coordinate system transformed to the mid-point of the rod:

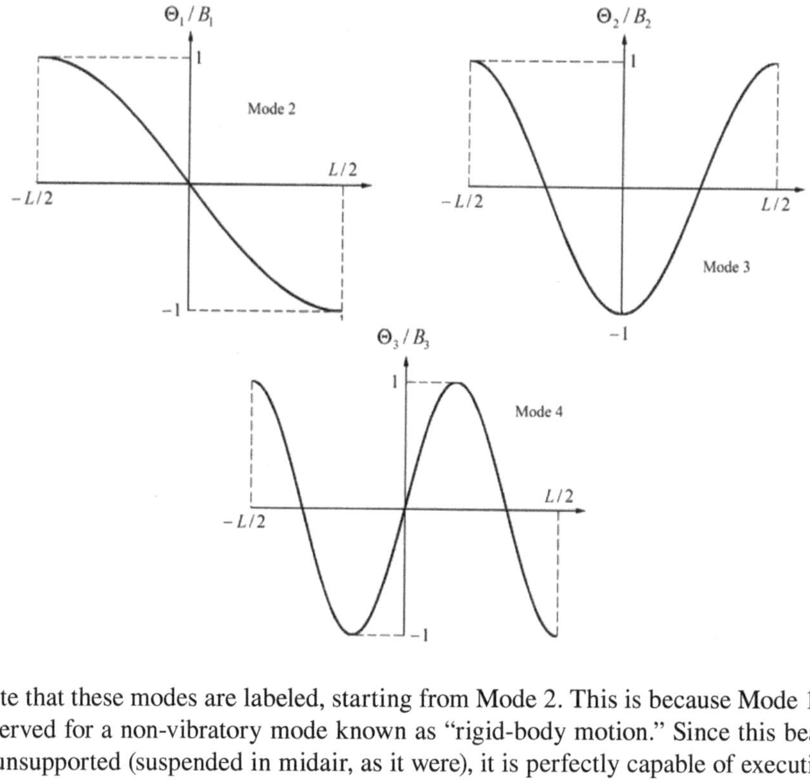

Note that these modes are labeled, starting from Mode 2. This is because Mode 1 is reserved for a non-vibratory mode known as "rigid-body motion." Since this beam is unsupported (suspended in midair, as it were), it is perfectly capable of executing a steady-state rotation about its longitudinal axis (y axis). One may solve for this by setting $\omega = 0$ in the differential equation for Θ, thus obtaining

$$\frac{d^2\Theta}{dy^2} = 0 \rightarrow \Theta = \hat{A} + \hat{B}y$$

The boundary conditions require that $\hat{B} = 0$, so one is left with Mode 1 as being

$$\left(\frac{\Theta}{\hat{A}}\right)_{Mode\,1} = 1$$

where $\hat{A}$ is the arbitrary amplitude.

5.3 BENDING VIBRATION

A segment of beam is shown as follows, upon which is imposed a distributed load $f(y)$:

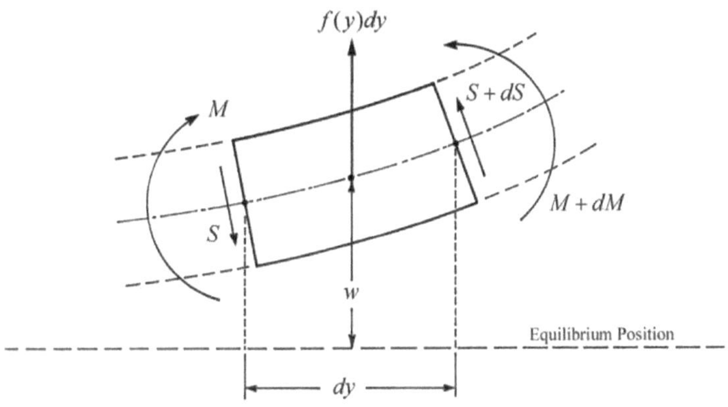

The force equilibrium equation is

$$S = S + dS + f(y)dy \rightarrow dS = -f(y)dy \rightarrow \frac{dS}{dy} = -f(y) \tag{5.7}$$

and the moment equilibrium equation is

$$\begin{aligned} M &= M + dM + (S + dS)dy + f(y)dy\frac{dy}{2} \rightarrow dM \\ &= -(S + dS)dy \rightarrow dM = -S\,dy \rightarrow S = -\frac{dM}{dy} \end{aligned} \tag{5.8}$$

Equations (5.7) and (5.8) combine to give

$$f(y) = \frac{d^2M}{dy^2} \tag{5.9}$$

Now, recall the elastic beam equation:

$$EI\frac{d^2w}{dy^2} = M(y)$$

Upon substituting this into Equation (5.9), one obtains

$$\frac{d^2}{dy^2}\left(EI\frac{d^2w}{dy^2}\right) = f(y)$$

Further, let f also be a function of time, $f(y,t)$, which is due to the inertial loading per unit length:

$$f(y,t) = -m\ddot{w}$$

where m is the mass per unit length. The equation becomes

$$\frac{\partial^2}{\partial y^2}\left(EI\frac{\partial^2 w}{\partial y^2}\right) = -m\ddot{w} \tag{5.10}$$

Choose a special case where $EI = Const.$ and $m = Const.$ Equation (5.10) takes the form

$$EI\frac{\partial^4 w}{\partial y^4} = -m\ddot{w}$$

Since this is now a linear differential equation, one may solve by the method of separation of variables:

$$w(y,t) = W(y)\,func(t) = W(y)\ \sin\omega t \tag{5.11}$$

Equation (5.11) into the previous equation gives a linear ordinary differential equation for $W(y)$:

$$\frac{d^4W}{dy^4} = \frac{m\omega^2}{EI}W \equiv k^4W \tag{5.12}$$

which has the general solution

$$W(y) = A\sin(ky) + B\cos(ky) + C\sinh(ky) + D\cosh(ky) \tag{5.13}$$

where A, B, C, and D are constants.

5.3.1 End Conditions for Beams

End conditions must now be imposed for the solution to proceed. The following figure shows the most common end conditions for beams (an example of a more complex end condition would be viscoelastic support):

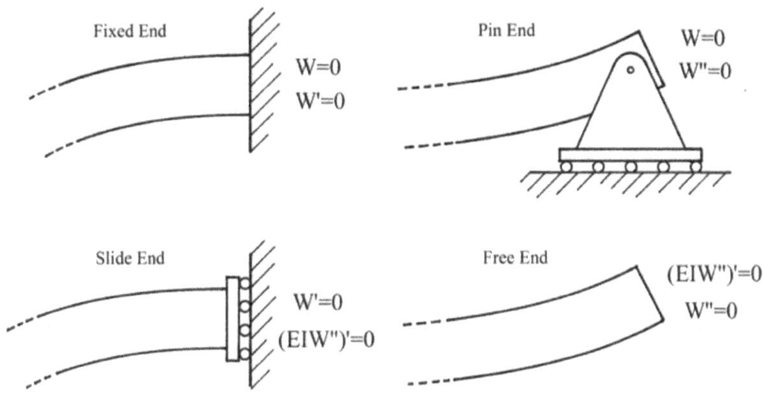

$W = 0$: zero displacement; $W' = 0$: zero slope

$W'' = 0$: zero bending moment; $\left(EIW''\right) = 0$: zero shear

5.3.2 Solution for a Uniform Cantilever Beam

The end conditions for a cantilever beam are applied to Equation (5.13). At $y = 0$, one has

$$W(0) = 0 \rightarrow B = -D;\ W'(0) = 0 \rightarrow A = -C$$

so that Equation (5.13) becomes

$$W(y) = C\left(\sinh(ky) - \sin(ky)\right) + D\left(\cosh(ky) - \cos(ky)\right) \tag{5.14}$$

At $y = L$, one has that $W''(L) = 0$, which, into Equation (5.14), gives

$$C\left(\sinh(kL) + \sin(kL)\right) + D\left(\cosh(kL) + \cos(kL)\right) = 0 \tag{5.15}$$

Also, at $y = L$, one has that $W'''(L) = 0$, which gives

$$C\left(\cosh(kL) + \cos(kL)\right) + D\left(\sinh(kL) - \sin(kL)\right) = 0 \tag{5.16}$$

With Equations (5.15) and (5.16), one either has the trivial solution that $C, D = 0$ or

$$\begin{vmatrix} (\sinh(kL) + \sin(kL) & (\cosh(kL) + \cos(kL) \\ (\cosh(kL) + \cos(kL) & (\sinh(kL) - \sin(kL) \end{vmatrix} = 0$$

When multiplied out, the characteristic equation is obtained:

$$\cos(kL)\ \cosh(kL) + 1 = 0$$

This may be solved for numerically:

$$\cos(kL) = -\frac{1}{\cosh(kL)}; \text{ Define: } \tilde{A} \equiv \cos(kL) \text{ and } \tilde{B} \equiv \frac{-1}{\cosh(kL)}$$

When $\tilde{A}$ and $\tilde{B}$ are plotted vs. kL, the intersection points where $\tilde{A} = \tilde{B}$ give the solutions.

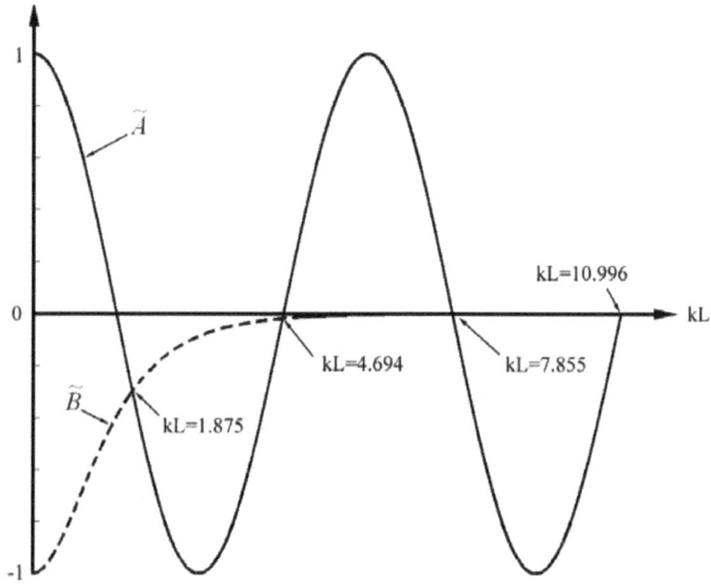

From this procedure, one obtains that

$$k_1L = 1.8751;\ k_2L = 4.6941;\ k_3L = 7.8548;\ k_4L = 10.9955;$$

Define

$$a_i \equiv (k_iL)^2$$

From the definition of k in Equation (5.12), the dynamic frequencies are

$$\omega_i = \frac{a_i}{L^2}\left(\frac{EI}{m}\right)^{1/2}$$

The mode shapes are found by solving Equation (5.15) for C and substituting this into Equation (5.14):

$$W_i(y) = D_i\left[\left(\cosh(k_iy) - \cos(k_iy)\right) + \left(\frac{\cosh(k_iL) + \cos(k_iL)}{\sinh(k_iL) + \sin(k_iL)}\right)\left(\sin(k_iy) - \sinh(k_iy)\right)\right]$$

The first three modes are plotted as follows:

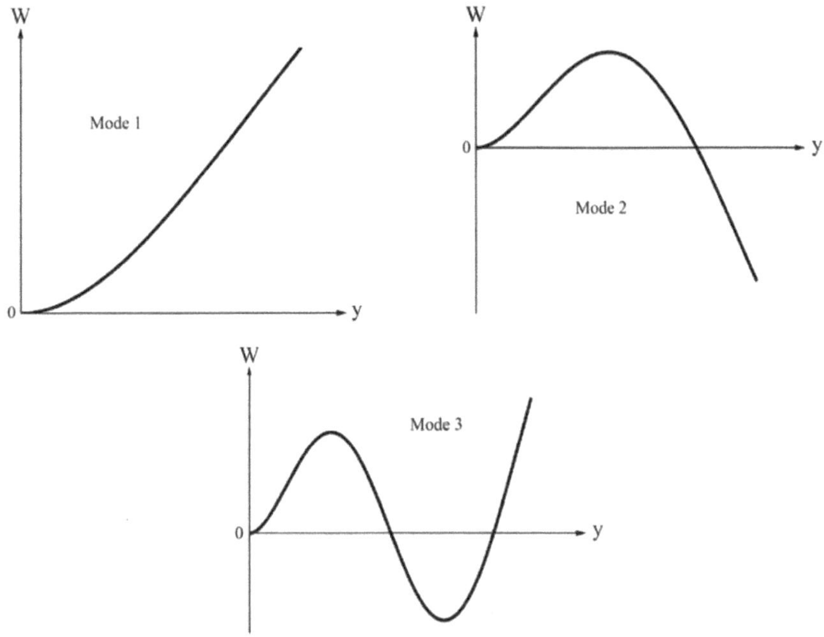

5.3.3 Solution for a Uniform Free-Free Beam

The end conditions for a free-free beam are now applied to Equation (5.13). At $y = 0$, one has

$$W''(0) = 0 \rightarrow B = D;\ W'''(0) = 0 \rightarrow A = C$$

and Equation (5.13) becomes

$$W(y) = C\big(\sin(ky) + \sinh(ky)\big) + D\big(\cos(ky) + \cosh(ky)\big) \tag{5.17}$$

At $y = L$, one has that $W''(L) = 0$, which, into Equation (5.17), gives

$$C\big(-\sin(kL) + \sinh(kL)\big) + D\big(-\cos(kL) + \cosh(kL)\big) = 0 \tag{5.18}$$

Further, $W'''(L) = 0$ into Equation (5.17) gives

$$C\big(-\cos(kL) + \cosh(kL)\big) + D\big(\sin(kL) + \sinh(kL)\big) = 0 \tag{5.19}$$

From Equations (5.18) and (5.19), one has either that $C, D = 0$, which is a trivial solution, or

$$\begin{vmatrix} \left(-\sin(kL)+\sinh(kL)\right) & \left(-\cos(kL)+\cosh(kL)\right) \\ \left(-\cos(kL)+\cosh(kL)\right) & \left(\sin(kL)+\sinh(kL)\right) \end{vmatrix} = 0 = 2\cos(kL)\cosh(kL)-2$$

Therefore, the characteristic equation is

$$\cos(kL)\cosh(kL)=1$$

which must be solved for numerically:

$$\cos(kL)=\frac{1}{\cosh(kL)}; \text{ Define: } \tilde{A}\equiv\cos(kL) \text{ and } \tilde{B}\equiv\frac{1}{\cosh(kL)}$$

Plot $\tilde{A}$ and $\tilde{B}$ vs. kL, and the intersection points (where $\tilde{A}=\tilde{B}$) give the solutions:

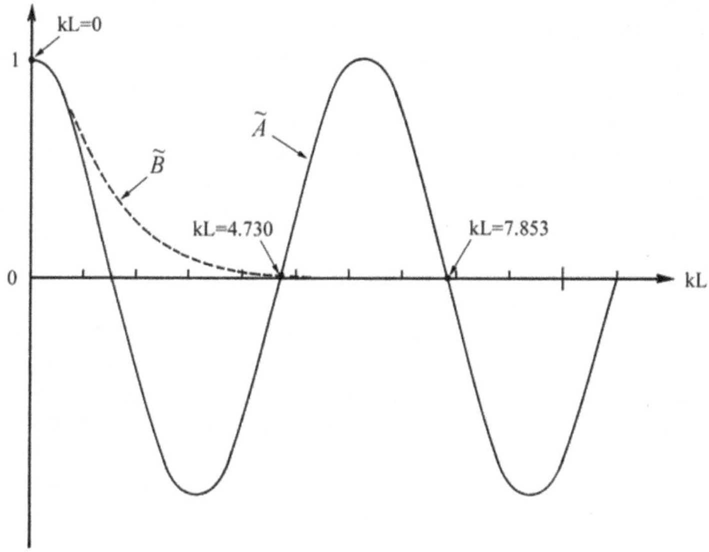

It can be seen that beyond the first solution point, all others are essentially the intersections of $\tilde{A}$ with the kL axis.

Now, from the previous definition,

$$a_i \equiv (k_i L)^2$$

and from the definition of k in Equation (5.12), the frequencies of vibration are given as before:

$$\omega_i = \frac{a_i}{L^2}\left(\frac{EI}{m}\right)^{1/2}$$

a_1 and a_2 are reserved for the so-called "rigid-body" modes, which will be defined later. For the vibratory modes, the solution gives that $a_3 = 22.373$, $a_4 = 61.673$, $a_5 = 120.903$, $a_6 = 199.860$, etc.

Equation (5.18) into Equation (5.17) gives the vibratory mode shapes (for $n \succ 2$):

$$W_i(y) = D_i \left[\left(\frac{\cos(k_i L) - \cosh(k_i L)}{\sinh(k_i L) - \sin(k_i L)} \right) \left(\sin(k_i y) + \sinh(k_i y) \right) + \left(\cosh(k_i y) + \cos(k_i y) \right) \right]$$

The first three modes plot out as shown (with the origin of the y coordinate transformed to mid-beam):

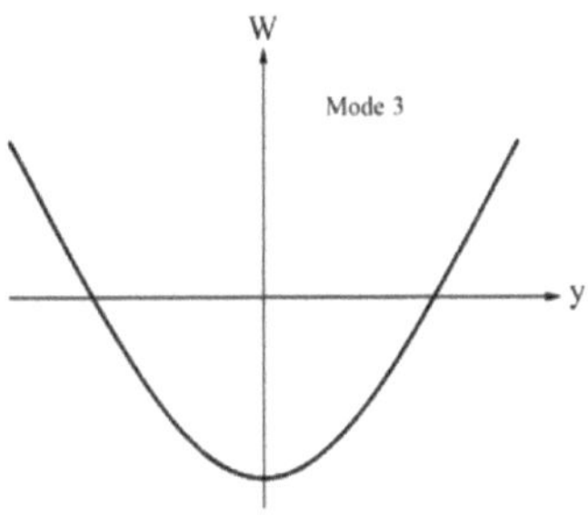

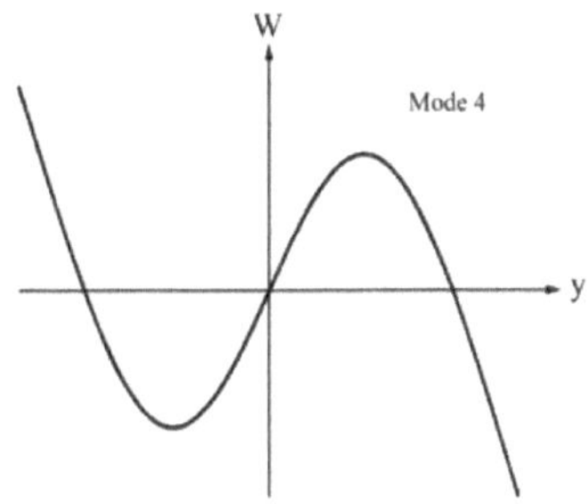

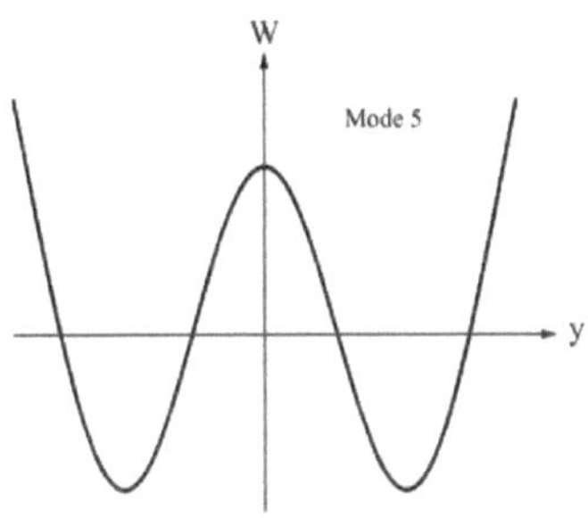

Now, if one substitutes $a_1 = 0$ and $a_2 = 0$ (k_1 and $k_2 = 0$) into the previous mode shape equation, one obtains

$$\frac{\sin(k_i y) + \sinh(k_i y)}{\sin(k_i L) - \sinh(k_i L)} = \frac{0}{0} \rightarrow \text{Indeterminant}$$

Therefore, the rigid-body mode shapes must be obtained by a different approach. Beginning with the fact that $\omega = 0$, Equation (5.12) becomes

$$\frac{d^4 W}{dy^4} = 0$$

which, upon integration, gives

$$W(y) = \hat{A}y^3 + \hat{B}y^2 + \hat{C}y + \hat{D}$$

The end conditions give that $\hat{A} = \hat{B} = 0$, so that the solution becomes

$$W(y) = \hat{C}y + \hat{D}$$

If one transforms the origin of the coordinate system to the center of the beam, this may be interpreted as the sum of the two rigid-body modes,

$$W_1(y) = \hat{C}y, \text{ rigid-body roll; } W_2(y) = \hat{D}, \text{ rigid-body heave}$$

and the shapes are shown:

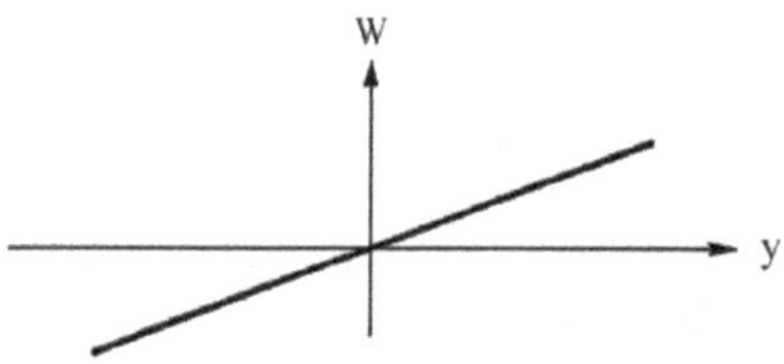

rigid-body roll

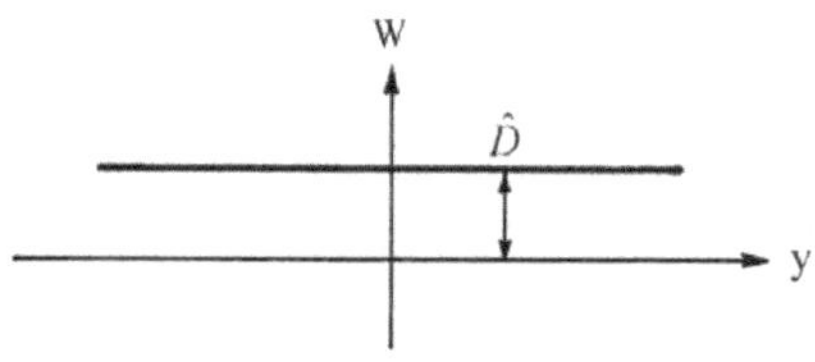

rigid-body heave

To those used to ground-fixed vibrating structures, where all modes are oscillatory, it may seem strange to define non-oscillating modes. However, this is a bit of "book-keeping" necessary for the non-supported free-free beam, where applied forces can give rise to linear and rotational accelerations and velocities. Also, note that the free-free beam is an aerospace-type problem in that it can represent slender space structures or wings in flight.

5.3.4 Separate Symmetrical and Anti-Symmetrical Solutions

An alternative way to solve for the free-free beam problem is to divide it into two half-beam problems, for symmetrical and anti-symmetrical modes:

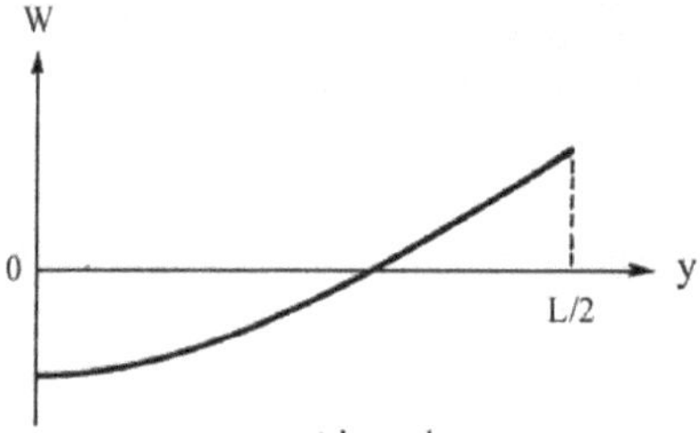

symmetric mode

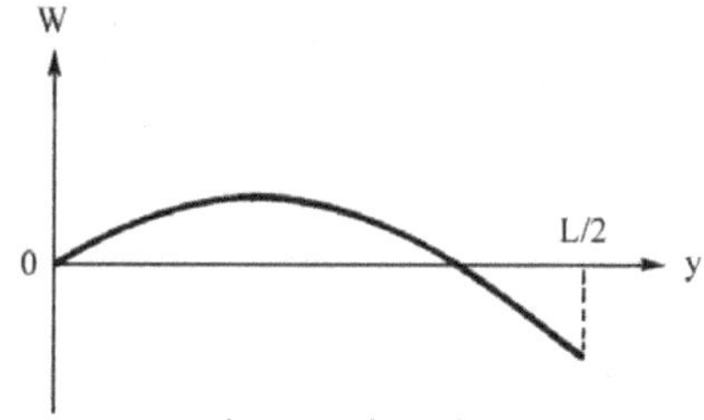

anti-symmetric mode

For the symmetric case, the end conditions at the root are

$$W'(0) = 0, \text{ (zero slope at root); } \left(EIW''\right)'(0) = 0, \text{ (zero shear at root)}$$

and for the anti-symmetric case, the root end conditions are

$$W(0) = 0, \text{ (zero displacement at root)}, \; W''(0) = 0, \text{ (pin restraint at root)}$$

For both cases, the $L/2$ end is free:

$$W''(L/2) = 0, \left(EIW''\right)'(L/2) = 0$$

5.3.5 Orthogonality Relations between Modes

Recall the beam's differential equation:

$$\left(EI\, w''\right)'' = -m\ddot{w}$$

Upon assuming a solution of the form $w = W(y)\sin \omega t$, one obtains

$$\left(EI\, W''\right)'' = m\omega^2 W \tag{5.20}$$

This gives the modes and frequencies:

$$W_r(y), \omega_r, r = 1, 2, 3, \ldots, \infty$$

Assume that these are distinct for each value of r, and substitute W_i, ω_i, and W_j, ω_j into Equation (5.20). This gives

$$\left(EI\, W_i''\right)'' = m\omega_i^2 W_i \tag{5.21}$$

$$\left(EI\, W_j\right)'' = m\omega_j^2 W_j \tag{5.22}$$

Multiply Equation (5.21) by W_j and (5.22) by W_i, and integrate from 0 to L:

$$\int_0^L \left(EI\, W_i\right) W_j \, dy = \omega_i^2 \int_0^L m\, W_i W_j \, dy \tag{5.23}$$

$$\int_0^L \left(EI\, W_j''\right)'' W_i \, dy = \omega_j^2 \int_0^L m\, W_i W_j \, dy \tag{5.24}$$

Equation (5.23) minus (5.24) gives

$$\left(\omega_i^2 - \omega_j^2\right) \int_0^L m\, W_i W_j \, dy = \int_0^L \left[\left(EI\, W_i''\right)'' W_j - \left(EI\, W_j''\right)'' W_i \right] dy \tag{5.25}$$

This may be integrated by parts, twice:

$$
\begin{aligned}
RHS = &\left[(EI\,W_i'')'W_j - (EI\,W_j'')'W_i\right]_0^L - \textstyle\int_0^L \left[(EI\,W_i'')W_j' - \right. \\
&\left.(EI\,W_j'')'\,W_i'\right]dy = \left[(EI\,W_i'')'W_j - (EI\,W_j'')'\,W_i\right]_0^L - \left[(EI\,W_i'')\,W_i' - \right. \\
&\left.(EI\,W_j'')W_i'\right]_0^L + \textstyle\int_0^L \left[(EI\,W_j'')W_j'' - (EI\,W_j'')W_i''\right]dy
\end{aligned}
\tag{5.26}
$$

Note that every single term in (5.26), including the integrand, is zero for all four end conditions previously described. For example, consider the free-free beam. Recall that the end conditions, at 0 and L, are $\left(EI\,W''\right) = 0$ and $W'' = 0$. When these are substituted into (5.26), every term becomes zero. Therefore, if $i \neq j$, then $\left(\omega_i^2 - \omega_j^2\right) \neq 0$, and Equation (5.25) gives

$$
\int_0^L m\,W_i W_j\,dy = 0 \tag{5.27}
$$

This is the orthogonality condition for vibrating beams. The two mode shapes are said to be orthogonal with respect to the mass distribution. Define

$$
M_{ij} \equiv \int_0^L m\,W_i W_j\,dy \tag{5.28}
$$

Therefore, orthogonality may be stated by

$$
M_{ij} = 0 \text{ for } i \neq j, \text{ and } M_{ii} = \int_0^L m\,W_i^2\,dy \succ 0 \text{ for } i = j
$$

M_{ii} is called the *generalized mass* associated with the ith mode of vibration. Its magnitude depends on how W_i is normalized.

Also, one may use Equation (5.24) to define orthogonality with respect to the stiffness distribution:

$$
K_{ij} \equiv \int_0^L \left(EI\,W_j''\right)''\,W_i\,dy = \omega_j^2 \int_0^L m\,W_i W_j\,dy
$$

From Equations (5.27) and (5.28), one obtains

$$
K_{ij} = 0 \text{ for } i \neq j, \text{ and } K_{ii} = \omega_i^2 M_{ii} \succ 0 \text{ for } i = j
$$

K_{ii} is called the *generalized stiffness* associated with the ith mode of vibration. Note that this development applies to *all* vibrating beams.

Now, upon returning to the uniform free-free beam, the orthogonality of the rigid modes will be assessed. Also, the coordinate system will be transformed so that $y = 0$ at the center of the beam. Note that there are four types of modes:

Rigid-body heave, $W_H = 1$
Rigid-body roll, $W_R = y/L$
Symmetric vibrating modes, W_S
Anti-symmetric vibrating modes, W_A

From Equation (5.28), one obtains:

$$M_{HH} = \int_{-L/2}^{L/2} m\,dy = M_{total} \text{ (total mass)}$$

$$M_{RR} = \int_{-L/2}^{L/2} m\left(y/L\right)^2 dy = I_{roll}/L^2 \text{ (total roll inertia / L}^2\text{)}$$

$$M_{HR}, M_{RH} = \int_{-L/2}^{L/2} m\left(y/L\right)1dy = 0 \text{ (odd function)}$$

Therefore, the rigid-body modes are orthogonal with respect to each other. Further, for the rigid-body modes in combination with the vibratory modes, one finds, first of all, that

$$M_{HA} = \int_{-L/2}^{L/2} m1W_A\,dy = 0 \text{ (odd function)}$$

$$M_{RS} = \int_{-L/2}^{L/2} m\left(y/L\right)W_S\,dy = 0 \text{ (odd function)}$$

Next, consider the beam in one of its vibratory symmetric or anti-symmetric modes:

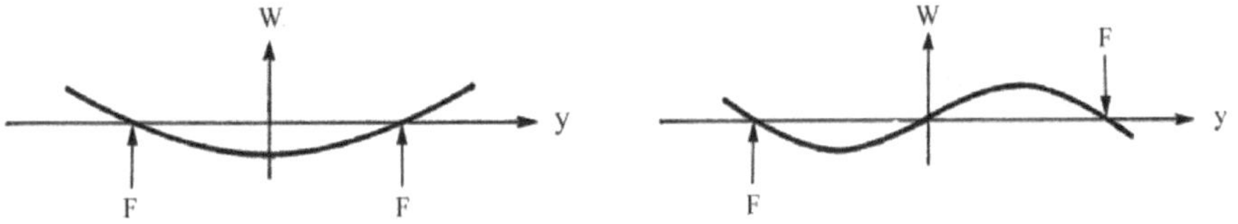

For the symmetric case, the vertical restraining forces at the node points are

$$F = \frac{1}{2}\int_{-L/2}^{L/2} m\,\ddot{w}\,dy = -\omega^2 \sin\omega t \int_{-L/2}^{L/2} mW_S\,dy$$

However, the free-free beam is assumed to have no external restraining forces, so

$$F \equiv 0 \rightarrow \int_{-L/2}^{L/2} mW_S\,dy = M_{HS} = 0$$

The same argument may be used for the anti-symmetric case, where the vertical restraining moment is

$$2FL_{node} = \int_{-L/2}^{L/2} m\,\ddot{w}\,y\,dy = -\omega^2 \sin\,\omega t \int_{-L/2}^{L/2} mW_A\;y\,dy = 0$$

which gives that

$$\int_{-L/2}^{L/2} mW_A\;y\,dy = 0 \rightarrow M_{RA} = \int_{-L/2}^{L/2} mW_A\left(y/L\right)dy = 0$$

Finally, as before, the vibratory modes are orthogonal to each other:

$$M_{AS} = \int_{-L/2}^{L/2} mW_A\,W_S\,dy = 0$$

Thus, all the modes of the free-free beam are orthogonal with respect to the mass distribution.

5.3.6 The "Work-Done" Interpretation

There is an interpretation of orthogonality in terms of the work done by the inertial reaction force.

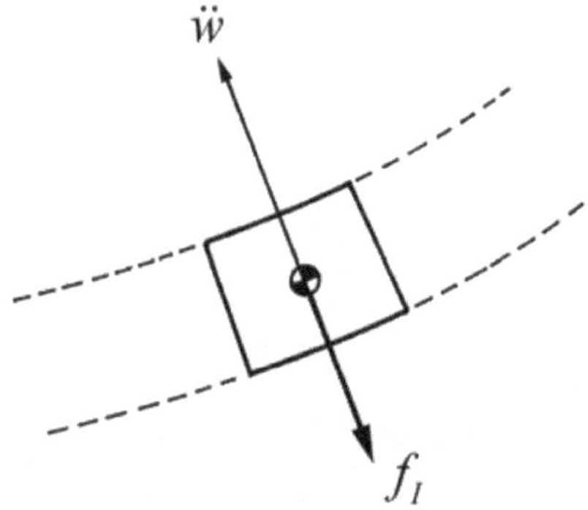

On a single beam element, the inertial reaction force is

$$f_I = -m\,\ddot{w}\,dy$$

For mode j, one has from before that

$$w_j = W_j \sin \omega_j t$$

Therefore,

$$f_I = mW_j\left(\omega_j^2 \sin \omega_j t\right)dy$$

One sees that mW_j is proportional to the magnitude of the inertial force per unit ω_j^2. Now, the work done by f_I when vibrating in mode i is

$$dW = f_I\, w_i = \omega_j^2 \sin \omega_j t \sin \omega_i t\left(mW_iW_j\right)dy$$

and the net work done on mode i by the inertial reaction force in mode j is

$$Work_{ij} = \omega_j^2 \sin \omega_j t \sin \omega_i t \int_{-L/2}^{L/2} mW_iW_j\, dy = \omega_j^2 \sin \omega_j t \sin \omega_i t\, M_{ij}$$

Thus, M_{ij} is the magnitude of the work done in mode i by the inertial reaction forces in mode j, per unit ω_j^2. The fact that $M_{ij} = 0$ when $i \neq j$ means that the reaction forces from mode j do *no* net work from the simultaneous motions of mode i.

One may perform a similar exercise to show that K_{ij} may be interpreted as the magnitude of the work done in mode i by the elastic forces in mode j.

5.4 APPROXIMATE METHODS FOR MODAL SOLUTIONS OF BENDING AND TORSION

5.4.1 Assumed Mode (or Rayleigh-Ritz) Method

Assume the deformation shape to be approximated by a linear combination of n functions $Z_i(y)$ that satisfy the end conditions. Therefore, the beam's deformation is given by

$$w(y,t)=\sum_{i=1}^{n} Z_i(y)q_{z_i}(t) \tag{5.29}$$

The $Z_i(y)$ functions are chosen to give a good approximation to the desired solution and must match the end conditions. The time-varying terms, $q_{z_i}(t)$, contain the generalized coordinates to be determined.

An element of the vibrating beam experiences variations in translational and rotational kinetic energy.

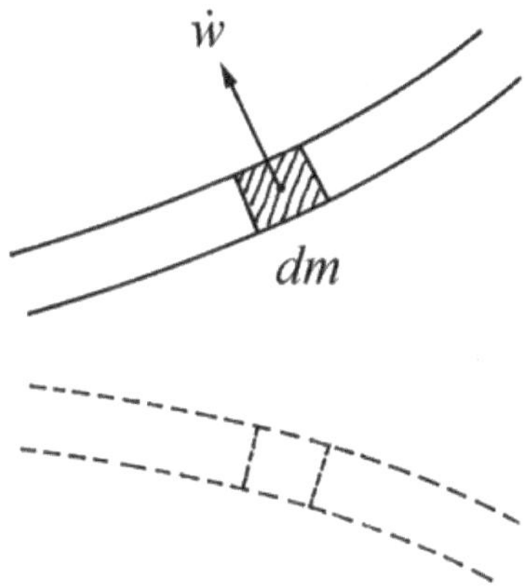

In this case, ignore rotational energy so that the kinetic energy is given by

$$T=\frac{1}{2}\int_0^L m(y)\dot{w}^2 dy \tag{5.30}$$

Also, from earlier work (Section 2.3), the strain energy is

$$V=\frac{1}{2}\int_0^L EI(w'')^2 dy \tag{5.31}$$

Equation (5.29) into (5.30) gives

$$T=\frac{1}{2}\int_0^L m\left(\sum_{i=1}^{n} Z_i\dot{q}_{z_i}\right)^2 dy=\frac{1}{2}\sum_{i=1}^{n}\sum_{j=1}^{n}\dot{q}_{z_i}\dot{q}_{z_j}\int_0^L mZ_iZ_j\,dy$$

$$=\frac{1}{2}\sum_{i=1}^{n}\sum_{j=1}^{n} M_{z_{ij}}\dot{q}_{z_i}\dot{q}_{z_j} \tag{5.32}$$

Similarly, Equation (5.29) into (5.31) gives

$$V = \frac{1}{2}\sum_{i=1}^{n}\sum_{j=1}^{n} K_{z_{ij}}\, q_{z_i}\, q_{z_j}, \text{ where } K_{z_{ij}} = \int_0^L EI\, Z_i Z_j'' dy \tag{5.33}$$

Now, Lagrange's equation, for no external forces on the system, is

$$\frac{d}{dt}\left(\frac{\partial T}{\partial \dot{q}_{z_i}}\right) + \frac{\partial V}{\partial q_{z_i}} = 0 \tag{5.34}$$

When Equations (5.32) and (5.33) are substituted into (5.34), one obtains

$$\sum_{j=1}^{n} M_{z_{ij}}\, \ddot{q}_{z_j} + \sum_{j=1}^{n} K_{z_{ij}}\, q_{z_j} = 0, 1 = 1,2,3,\dots\dots,n \tag{5.35}$$

If one assumes a solution of the form

$$q_{z_j} = \overline{q}_{z_j} \sin \omega t$$

then Equation (5.35) becomes

$$\sum_{j=1}^{n}\left(K_{z_{ij}}\, \overline{q}_{z_j} - M_{z_{ij}}\, \omega^2\, \overline{q}_{z_j}\right) = 0$$

This may also be expressed in matrix form:

$$\left[K_z - \omega^2 M_z\right]\left\{\overline{q}_z\right\} = 0$$

The eigenvalues of this give the modal frequencies, ω_i, and the eigenvectors, $\left\{\overline{q}_z\right\}$, give the magnitudes of the assumed modes.

The same methodology may be used to obtain approximate solutions for torsional vibration. In this case, the torsional deformation is given by

$$\theta = \sum_{i=1}^{n} \Theta_i(y)\, q_{\theta_i}(t)$$

and the kinetic and elastic strain energies are

$$T = \frac{1}{2}\int_0^L i_{cm}\, \dot{\theta}^2 dy, V = \frac{1}{2}\int_0^L GJ\left(\theta'\right)^2 dy$$

where i_{cm} is the rotational moment of inertia about the elastic axis (which is assumed to be coincident with the cross-sectional mass center). With the substitution of the assumed modes, these energy expressions become

$$T = \frac{1}{2}\sum_{i=1}^{n}\sum_{j=1}^{n} M_{\theta_{ij}}\, \dot{q}_{\theta_i}\, \dot{q}_{\theta_j}, \text{ where } M_{\theta_{ij}} \equiv \int_0^L i_{cm}\, \Theta_i\, \Theta_j\, dy$$

$$V = \frac{1}{2}\sum_{i=1}^{n}\sum_{j=1}^{n} K_{\theta_{ij}}\, q_{\theta_i}\, q_{\theta_j}, \text{ where } K_{\theta_{ij}} = \int_0^L GJ\,\Theta_i'\,\Theta_j'\, dy$$

Upon application of Lagrange's equation, as before, the dynamic equations are obtained:

$$\sum_{j=1}^{n} M_{\theta_{ij}}\, \ddot{q}_{\theta_j} + \sum_{j=1}^{n} K_{\theta_{ij}}\, q_{\theta_j} = 0, 1 = 1,2,3,....,n$$

The assumed solution is

$$q_{\theta_j} = \overline{q}_{\theta_j} \sin \omega t$$

which gives

$$\sum_{j=1}^{n}\left(K_{\theta_{ij}}\, \overline{q}_{\theta_j} - M_{\theta_{ij}}\, \omega^2\, \overline{q}_{\theta_j}\right) = 0 \rightarrow \left[K_\theta - \omega^2 M_\theta\right]\{\overline{q}_\theta\} = 0$$

As for the bending case, the eigenvalues give the modal frequencies, ω_i, and the eigenvectors, $\{\overline{q}_\theta\}$, give the magnitudes of the assumed modes.

5.4.1.1 Example 5.1: Bending Vibration

Consider an example *non-uniform* beam with the following properties:

$$L = 6.096m;\ EI = 9.7523\times 10^6\left[1-0.9(y/L)^{0.25}\right];\ m = 35.7087\left[1-0.8(y/L)^{0.5}\right]$$

For the assumed modes, choose the first three modes for the *uniform* cantilever beam from Section 5.3.2:

$$\begin{aligned} Z_i(y) &= \cosh\left[(k_iL)y/L\right] - \cos\left[(k_iL)y/L)\right] \\ &\quad + \beta_i\left\{\sin\left[(k_iL)y/L)\right] - \sinh\left[(k_iL)y/L)\right]\right\} \end{aligned}$$

where

$$k_1L = 1.8751,\ k_2L = 4.6941,\ k_3L = 7.8548,\ \beta_1 = 0.7341,\\ \beta_2 = 1.0185,\ \beta_3 = 0.9992$$

The $M_{z_{ij}}$ terms may now be calculated from

$$M_{z_{ij}} = \int_0^L m Z_i Z_j\, dy$$

which gives the $\left[M_z\right]$ matrix as

$$\left[M_z\right] = \begin{bmatrix} 62.1275 & 16.1333 & -0.4595 \\ 16.1333 & 86.4173 & 22.6126 \\ -0.4595 & 22.6126 & 95.3783 \end{bmatrix}$$

The second derivative of the modes is

$$Z_i''(y) = k_i^2 \left\langle \cosh\left[(k_i L) y / L\right] + \cos\left[(k_i L) y / L)\right] \right.$$
$$\left. -\beta_i \left\{ \sin\left[(k_i L) y / L)\right] + \sinh\left[(k_i L) y / L)\right] \right\} \right\rangle$$

From which one may obtain the $K_{z_{ij}}$ terms and $\left[K_z\right]$ matrix:

$$K_{z_{ij}} = \int_0^L EI\, Z_i'' Z_j''\, dy \rightarrow \left[K_z\right] = \begin{bmatrix} 0.0238 & 0.0398 & 0.0476 \\ 0.0398 & 0.7001 & 0.6642 \\ 0.0476 & 0.6642 & 5.0150 \end{bmatrix} \times 10^7$$

The resulting eigenvalue solution gives frequencies of

$$\omega_1 = 59.559,\ \omega_2 = 276.686,\ \omega_3 = 731.385\ (rad / \sec)$$

and the corresponding eigenvectors are

$$\{\bar{q}_z\}_1 = \begin{Bmatrix} 0.9989 \\ -0.0476 \\ -0.0033 \end{Bmatrix},\ \{\bar{q}_z\}_2 = \begin{Bmatrix} 0.1930 \\ -0.9751 \\ 0.1094 \end{Bmatrix},\ \{\bar{q}_z\}_3 = \begin{Bmatrix} 0.0589 \\ -0.1496 \\ 0.9870 \end{Bmatrix}$$

Note that for $\{\bar{q}_z\}_1$, the dominant term is $\bar{q}_{z_1}$, which means that the mode shape is primarily that of the assumed first mode. $\bar{q}_{z_2}$ and $\bar{q}_{z_3}$ decrease progressively in magnitude, which confirms that the assumed mode was a good choice. The same results are seen for the other two eigenvectors. For example, the shape of the third eigenvector is dominated by the assumed third mode. However, the resulting solution is certainly not exactly the same as the assumed modes because of the non-uniformity of EI and m along the length of the beam.

5.4.1.2 Example 5.2: Torsional Vibration

Consider an example non-uniform torsion member with the following properties:

$$L = 6.096m;\ GJ = 9.8763 \times 10^5 \left[1 - 0.9 (y / L)^{0.25}\right];$$
$$i_{cm} = 7.4544 \left[1 - 0.8 (y / L)^{0.33}\right]$$

For the assumed modes, choose the first three modes for the *uniform* cantilever rod from Section 5.2.1 (A_j is arbitrarily chosen to be unity):

$$\Theta_i(y) = \sin[a_i (y / L)]$$

where

$$a_1 = 1.5708,\ a_2 = 4.7124,\ a_3 = 7.8540$$

The $M_{\theta_{ij}}$ terms may now be calculated from

$$M_{\theta_{ij}} \equiv \int_0^L i_{cm}\, \Theta_i \Theta_j \, dy$$

which gives the $[M_\theta]$ matrix as

$$[M_\theta] = \begin{bmatrix} 6.7342 & 1.7702 & 0.0760 \\ 1.7702 & 8.5803 & 2.0801 \\ 0.0760 & 2.0801 & 8.8236 \end{bmatrix}$$

The derivative of the modes is

$$\Theta_i'(y) = (a_i / L)\cos[a_i(y/L)]$$

From which one may obtain the $K_{\theta_{ij}}$ terms and $[K_\theta]$ matrix:

$$K_{\theta_{ij}} = \int_0^L GJ\, \Theta_i' \Theta_j' \, dy \rightarrow [K_\theta] = \begin{bmatrix} 0.0751 & 0.0729 & 0.0473 \\ 0.0729 & 0.5424 & 0.3222 \\ 0.0473 & 0.3222 & 1.4541 \end{bmatrix} \times 10^6$$

The resulting eigenvalue solution gives frequencies of

$$\omega_1 = 100.769,\ \omega_2 = 242.535,\ \omega_3 = 406.245\,(rad/\sec)$$

and the corresponding eigenvectors are

$$\{\bar{q}_\theta\}_1 = \begin{Bmatrix} -0.9934 \\ 0.1140 \\ 0.0087 \end{Bmatrix}, \{\bar{q}_\theta\}_2 = \begin{Bmatrix} -0.1216 \\ 0.9718 \\ -0.2021 \end{Bmatrix}, \{\bar{q}_\theta\}_3 = \begin{Bmatrix} 0.0408 \\ -0.0343 \\ 0.9986 \end{Bmatrix}$$

As for the bending example, the dominant term in $\{\bar{q}_\theta\}_1$ is $\bar{q}_{\theta_1}$, which means that the mode shape is primarily that of the assumed first mode. $\bar{q}_{\theta_2}$ and $\bar{q}_{\theta_3}$ decrease progressively in magnitude, which confirms that the assumed mode was a good choice. The same results are seen for the other two eigenvectors.

5.4.2 Assumed-Mode Method for Combined Bending and Twisting

If the elastic center and mass center of a beam were coincident, then bending and twisting dynamics would be uncoupled. However, for a typical wing, these centers are displaced from each other, and the motions are coupled.

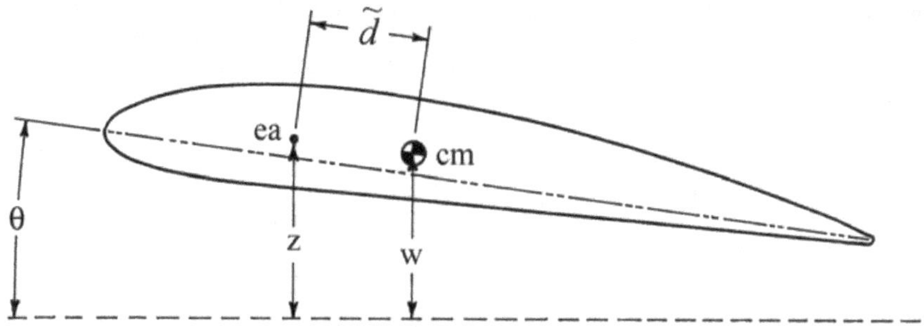

In this case, modes are assumed for both displacement and rotation:

$$z=\sum_{i=1}^{n} Z_i(y)q_{z_i}(t),\ \theta=\sum_{i=1}^{n}\Theta_i(y)q_{\theta_i}(t) \tag{5.36}$$

Also, the strain energy is due to bending and twisting:

$$V=V_{bending}+V_{twisting}=\frac{1}{2}\int_0^L EI\left(z''\right)^2 dy+\frac{1}{2}\int_0^L GJ\left(\theta'\right)^2 dy$$

Upon substituting Equations (5.36), this becomes

$$V=\frac{1}{2}\sum_{i=1}^{n}\sum_{j=1}^{n}\left(K_{z_{ij}}\,q_{z_i}\,q_{z_j}+K_{\theta_{ij}}\,q_{\theta_i}\,q_{\theta_j}\right) \tag{5.37}$$

where

$$K_{z_{ij}}=\int_0^L EI\,Z_i''Z_j''dy,\ K_{\theta_{ij}}=\int_0^L GJ\,\Theta_i'\Theta_j'\,dy$$

The kinetic energy also involves translation and rotation:

$$T=T_{\text{velocity of cm}}+T_{\text{rotation about cm}}=\frac{1}{2}\int_0^L m\,\dot{w}^2\,dy+\frac{1}{2}\int_0^L i_{cm}\,\dot{\theta}^2\,dy \tag{5.38}$$

Now, for small values of θ, $w=z-\theta\tilde{d}$. Also,

$$i_{cm}=i_{ea}-m\tilde{d}^2\equiv m\left(R^2-\tilde{d}^2\right)$$

where R is the section's mass radius of gyration about its elastic center. These equations into Equation (5.38) give

$$T=\frac{1}{2}\int_0^L\left[m\left(\dot{z}-\dot{\theta}\tilde{d}\right)^2+m\left(R^2-\tilde{d}^2\right)\dot{\theta}^2\right]dy=\frac{1}{2}\int_0^L m\left(\dot{z}^2-2\dot{z}\dot{\theta}\tilde{d}+R^2\dot{\theta}^2\right)dy$$

and Equations (5.36) into this gives

$$T=\frac{1}{2}\sum_{i=1}^{n}\sum_{j=1}^{n}\left(M_{zz_{ij}}\dot{q}_{z_i}\dot{q}_{z_j}+2M_{z\theta_{ij}}\dot{q}_{z_i}\dot{q}_{\theta_j}+M_{\theta\theta_{ij}}\dot{q}_{\theta_i}\dot{q}_{\theta_j}\right) \tag{5.39}$$

where

$$M_{zz_{ij}}\equiv\int_0^L m Z_i Z_j\,dy,\; M_{z\theta_{ij}}=M_{\theta z ji}\equiv-\int_0^L m\tilde{d}\Theta_i Z_j\,dy$$

$$M_{\theta\theta_{ij}}\equiv\int_0^L mR^2\Theta_i\Theta_j\,dy=\int_0^L i_{ea}\Theta_i\Theta_j\,dy$$

Lagrange's equations for an unforced system are

$$\frac{d}{dt}\left(\frac{\partial T}{\partial \dot{q}_{z_i}}\right)+\frac{\partial V}{\partial q_{z_i}}=0,\;\frac{d}{dt}\left(\frac{\partial T}{\partial \dot{q}_{\theta_i}}\right)+\frac{\partial V}{\partial q_{\theta_i}}=0 \tag{5.40}$$

Equations (5.37) and (5.39) into (5.40) give a system of second-order linear ordinary differential equations:

$$\sum_{j=1}^{n}\left(M_{zz_{ij}}\ddot{q}_{z_j}+M_{z\theta_{ij}}\ddot{q}_{\theta_j}+K_{z_{ij}}q_{z_j}\right)=0 \tag{5.41}$$

$$\sum_{j=1}^{n}\left(M_{\theta\theta_{ij}}\ddot{q}_{\theta_j}+M_{\theta z_{ij}}\ddot{q}_{z_j}+K_{\theta_{ij}}q_{\theta_j}\right)=0 \tag{5.42}$$

Now, assume the following solutions for these equations:

$$q_{z_j}=\overline{q}_{z_j}\sin\omega t,\; q_{\theta_j}=\overline{q}_{\theta_j}\sin\omega t$$

When substituted, Equations (5.41) and (5.42) become

$$\sum_{j=1}^{n}\left[K_{z_{ij}}\overline{q}_{z_j}-\omega^2\left(M_{zz_{ij}}\overline{q}_{z_j}+M_{z\theta_{ij}}\overline{q}_{\theta_j}\right)\right]=0 \tag{5.43}$$

$$\sum_{j=1}^{n}\left[K_{\theta_{ij}}\overline{q}_{\theta_j}-\omega^2\left(M_{\theta\theta_{ij}}\overline{q}_{\theta_j}+M_{\theta z_{ij}}\overline{q}_{z_j}\right)\right]=0 \tag{5.44}$$

Define the matrices:

$$[K]\equiv\begin{bmatrix}K_{z_{ij}} & 0\\ 0 & K_{\theta_{ij}}\end{bmatrix},\;[M]\equiv\begin{bmatrix}M_{zz_{ij}} & M_{z\theta_{ij}}\\ M_{\theta z_{ij}} & M_{\theta\theta_{ij}}\end{bmatrix},\;\{\overline{q}\}\equiv\begin{Bmatrix}\overline{q}_{z_j}\\ \overline{q}_{\theta_j}\end{Bmatrix}$$

With these, Equations (5.41) and (5.42) may be expressed as

$$\left[K-\omega^2 M\right]\{\bar{q}\}=0$$

As described before, the eigenvalues give the modal frequencies, ω_i, and the eigenvectors, $\{\bar{q}\}$, give the magnitudes of the assumed modes.

5.4.2.1 Example 5.3

A non-uniform beam has the following properties:

$$L=6.096m;\ EI=9.7523\times10^6\left[1-0.9(y/L)^{0.25}\right];$$

$$m=35.7087\left[1-0.8(y/L)^{0.5}\right]$$

$$GJ=9.8763\times10^5\left[1-0.9(y/L)^{0.25}\right];\ i_{cm}=7.4544\left[1-0.8(y/L)^{0.33}\right];$$

$$\tilde{d}=0.1826\left[1-0.5(y/L)\right]$$

Further, choose the assumed bending modes from Example 5.1 and torsion modes from Example 5.2. Therefore, the $K_{z_{ij}}$ and $K_{\theta_{ij}}$ terms are the same as before, and one has that

$$[K]=\begin{bmatrix} 2.3823 & 3.9828 & 4.7613 & 0 & 0 & 0 \\ 3.9828 & 70.012 & 66.417 & 0 & 0 & 0 \\ 4.7613 & 66.417 & 501.50 & 0 & 0 & 0 \\ 0 & 0 & 0 & 0.7509 & 0.7287 & 0.4732 \\ 0 & 0 & 0 & 0.7287 & 5.4237 & 3.2219 \\ 0 & 0 & 0 & 0.4732 & 3.2219 & 14.541 \end{bmatrix}\times10^5$$

Also, one sees that $M_{zz_{ij}}$ is the same as $M_{z_{ij}}$ in Example 5.1. $M_{\theta\theta_{ij}}$ differs from $M_{\theta_{ij}}$ of Example 5.2 because i_{ea} does not equal the i_{cm} for that example. The cross terms, $M_{z\theta_{ij}}$, are calculated from

$$M_{z\theta_{ij}}=M_{\theta z_{ij}}\equiv-\int_0^L m\tilde{d}\Theta_i Z_j\,dy=\begin{bmatrix} -5.3181 & -0.44709 & 0.53425 \\ -4.4525 & -7.9103 & 0.55316 \\ -1.6713 & -6.8536 & -8.0069 \end{bmatrix}$$

so that

$$[M]=\begin{bmatrix} 62.128 & 16.133 & -0.4595 & -5.3181 & -0.4471 & 0.5343 \\ 16.133 & 86.417 & 22.613 & -4.4525 & -7.9103 & 0.5532 \\ -0.4595 & 22.613 & 95.378 & -1.6713 & -6.8536 & -8.0069 \\ -5.3181 & -4.4525 & -1.6713 & 7.3182 & 2.1607 & 0.1360 \\ -0.4471 & -7.9103 & -6.8536 & 2.1607 & 9.6149 & 2.5772 \\ 0.5343 & 0.5532 & -8.0069 & 0.1360 & 2.5772 & 9.9151 \end{bmatrix}$$

The solution for the eigenvalue problem gives the following modal frequencies:

$$\omega_1 = 58.536, \omega_2 = 101.79, \omega_3 = 219.01, \omega_4 = 303.30$$

$$\omega_5 = 381.59, \omega_6 = 776.61\,(rad\,/\,\sec)$$

And the corresponding eigenvectors are the modal shapes, such as for the first mode:

$$\{\overline{q}\}_1 = \begin{Bmatrix} \{\overline{q}_z\}_1 \\ \{\overline{q}_\theta\}_1 \end{Bmatrix} = \begin{Bmatrix} 0.91996 \\ -0.04335 \\ -0.00309 \\ -0.38670 \\ 0.04721 \\ 0.00353 \end{Bmatrix}$$

This clearly shows the coupling between the bending and torsion dynamics. However, because the frequency is close to the value of 59.559 *rad*/sec for the uncoupled bending case and $\{\overline{q}_z\}_1$ shows the dominance of the first assumed bending mode, $\{\overline{q}\}_1$ may be interpreted as primarily a first mode bending shape, with some torsional coupling, of course. The second mode is

$$\{\overline{q}\}_2 = \begin{Bmatrix} \{\overline{q}_z\}_2 \\ \{\overline{q}_\theta\}_2 \end{Bmatrix} = \begin{Bmatrix} -0.12678 \\ 0.01084 \\ 0.00001 \\ -0.98624 \\ 0.10505 \\ 0.01006 \end{Bmatrix}$$

In this instance, the same reasoning as before leads to the conclusion that this is the coupled version of the first torsional mode (from Example 5.2). In summary, the eigenvectors for all six modes are

$$\{\overline{q}\} = \begin{bmatrix} 0.91996 & -0.12678 & 0.00829 & 0.08987 & 0.00487 & 0.03516 \\ -0.04335 & 0.01084 & -0.13119 & -0.48076 & -0.06165 & -0.08408 \\ -0.00309 & 0.00001 & 0.01108 & 0.06505 & -0.02129 & 0.68654 \\ -0.38670 & -0.98624 & -0.22437 & -0.05546 & 0.02845 & 0.04254 \\ 0.04721 & 0.10505 & 0.94657 & -0.86792 & -0.17598 & 0.30219 \\ 0.00353 & 0.01006 & -0.19043 & 0.01448 & 0.98181 & 0.65363 \end{bmatrix}$$

5.4.3 Lumped-Parameter Models

5.4.3.1 Flexibility-Matrix Formulation

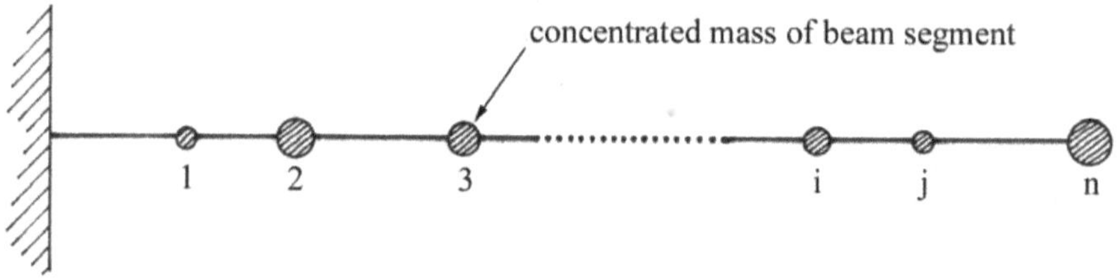

Consider the beam to be conceptually divided into elastic segments, with the mass of each segment concentrated at specific stations. Recall from previous work that the deformation of a continuous beam is given by

$$w(y) = \int_0^L C^{zz}(y,\eta) Z(\eta) d\eta \tag{5.45}$$

$Z(\eta)$ is the load per span, which, in this case, is the inertial reaction force:

$$Z(\eta) = -m(\eta)\ddot{w}(\eta)$$

If one assumes a solution of the form

$$w = W(y) \sin \omega t$$

then Equation (5.45) becomes

$$W(y) = \omega^2 \int_0^L C^{zz} m(\eta) W(\eta) d\eta \tag{5.46}$$

For the segmented beam, one may define

$$W(y) = W_i,\ W(\eta) = W_j,\ C^{zz}(y,\eta) = C_{ij}^{zz}$$

$$\int_{segment} m(\eta) d\eta = m_j = \text{mass of each segment}$$

Therefore, Equation (5.46) becomes

$$W_i = \omega^2 \sum_{j=1}^{n} C_{ij}^{zz} m_j W_j \qquad i = 1, 2, \ldots\ldots\ldots, n$$

This may also be expressed in matrix form:

$$\{W\} = \omega^2 \left[C^{zz}\right][m]\{W\} \tag{5.47}$$

where $[m]$ is a diagonal matrix.

One may define

$$[C^{zz}][m] \equiv [U] = \text{"}\textit{dynamic matrix}\text{"}, \lambda \equiv 1/\omega^2 = \text{"}\textit{frequency parameter}\text{"}$$

So Equation (5.47) becomes

$$[U - \lambda I]\{W\} = 0$$

which is an eigenvalue problem for the distributed magnitudes of each mode, $\{W\}$, and the associated frequency parameters, λ.

A similar exercise may be performed for the torsional modes of vibration. Starting from

$$\theta(y) = \int_0^L C^{\theta\theta}(y,\eta)\, t(\eta)\, d\eta$$

one may define the torsional load per unit span, $t(\eta)$, as

$$t(\eta) = -i_\theta(\eta)\ddot{\theta}(\eta)$$

where $i_\theta(\eta)$ is the torsional mass moment of inertia about the beam's elastic axis. Upon assuming a solution of the form

$$\theta = \Theta(y) \sin \omega t$$

one obtains that

$$\Theta(y) = \omega^2 \int_0^L C^{\theta\theta} i_\theta(\eta)\Theta(\eta)\, d\eta$$

For the segmented beam, this becomes

$$\Theta_i = \omega^2 \sum_{j=1}^{n} C_{ij}^{\theta\theta}(i_\theta)_j \Theta_j \quad i = 1, 2, \ldots\ldots., n$$

where

$$\Theta(y) = \Theta_i, \Theta(\eta) = \Theta_j, C^{\theta\theta}(y,\eta) = C_{ij}^{\theta\theta}$$

$$\int_{segment} i_\theta(\eta)\, d\eta = (i_\theta)_j = \text{mass moment of inertia of each segment}$$

Again, this may be cast into matrix form:

$$\{\Theta\} = \omega^2 \left[C^{\theta\theta}\right][i_\theta]\{\Theta\}$$

which gives the eigenvalue problem for the dynamic modes, $\{\Theta\}$, and frequency parameters, $\lambda = 1/\omega^2$:

$$\left[U_\theta - \lambda I\right]\{\Theta\} = 0$$

In this case, the "dynamic matrix" is

$$\left[U_\theta\right] = \left[C^{\theta\theta}\right]\left[i_\theta\right]$$

5.4.3.2 Example 5.4

Consider the segmented-beam representation of the wing in Example 2.4. Further, assume that the mass and inertial centers lie along the elastic axis. Therefore, the bending and twisting modes are uncoupled. The values of m and i_θ for each segment are:

Segment Number	m (kg)	$i_\theta \left(kg - m^2\right)$
1	2739.2	2551.82
2	4626.6	27262.29
3	1905.1	1088.62
4	1542.2	810.61
5	308.4	117.06

From the flexibility influence coefficients of Example 2.4 and the preceding eigenvalue equations for bending and twisting, one may obtain that:

Mode	$\omega_{bending}$ (rad / sec)	$\omega_{twisting}$ (rad / sec)
1	10.942	23.338
2	41.427	104.58
3	108.71	161.91
4	195.71	230.41
5	289.83	251.64

Further, the modal vectors for the first bending and twisting modes are

$$\{W\}_1 = \begin{Bmatrix} 0.0407 \\ 0.1565 \\ 0.3071 \\ 0.5385 \\ 0.7678 \end{Bmatrix}, \{\Theta\}_1 = \begin{Bmatrix} 0.2305 \\ 0.4662 \\ 0.4820 \\ 0.4967 \\ 0.5005 \end{Bmatrix}$$

For aerospace applications, there is a shortcoming to the flexibility formulation in that C^{zz} and $C^{\theta\theta}$ are not defined for an unsupported beam. Scanlan and

Rosenbaum describe a method to get around this in *Introduction to the Study of Aircraft Vibration and Flutter*, Macmillan Co., 1962, pp. 173–175. However, an alternative method based on the stiffness influence coefficient matrix is a more direct way to model an unsupported beam.

5.4.3.3 Stiffness Matrix Formulation

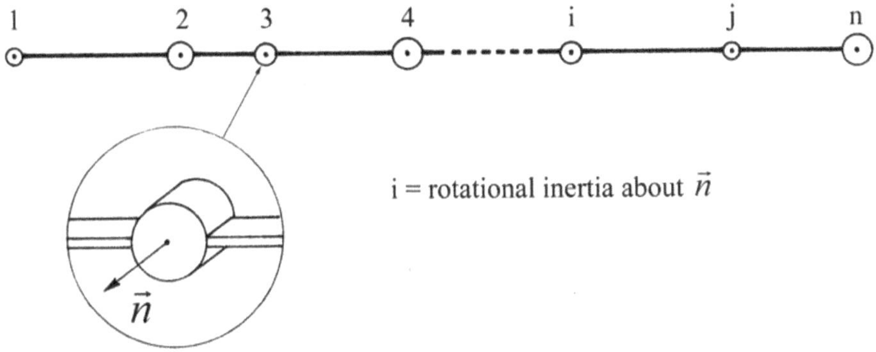

This segmented beam is assumed to be composed of massless uniform elastic elements between concentrated masses. The mass of the actual beam's segment is apportioned to the concentrated masses at either end. Note that these masses are shared between adjoining segments. This first step in constructing the stiffness matrix is to consider the case of a uniform cantilever beam:

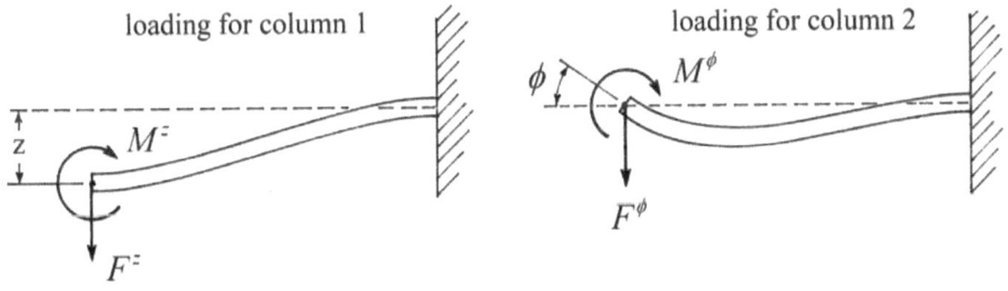

The coordinates of interest are the vertical displacement, z, and the slope, ϕ. The first column in the stiffness influence coefficient matrix is obtained by determining the force and moment, F^z and M^z, that will give a unit value of z and zero ϕ:

$$F^z = \frac{12EI}{L^3}, M^z = \frac{6EI}{L^2}$$

The second column is obtained by finding the force and moment, F^ϕ and M^ϕ, that will give a unit value of ϕ and zero z:

$$F^\phi = \frac{6EI}{L^2}, M^\phi = \frac{4EI}{L}$$

Therefore, the 2 × 2 matrix is

$$[K]=\begin{bmatrix}\dfrac{12EI}{L^3} & \dfrac{6EI}{L^2}\\ \dfrac{6EI}{L^2} & \dfrac{4EI}{L}\end{bmatrix}$$

Next, the stiffness matrix is found for a free beam segment. This may be obtained by replacing the wall for the cantilever beam with the reaction forces and moments that give equilibrium.

loading for column 1

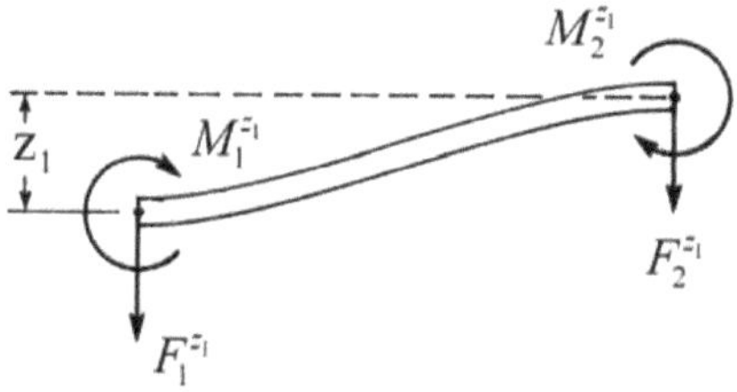

loading for column 2

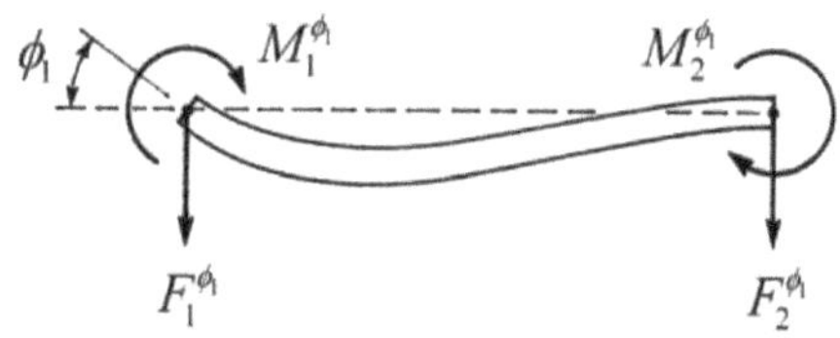

The value of the reaction forces and moments may be found from

$$F_2=-F_1,\, M_2=-M_1+F_1L$$

which gives

$$F_1^{z_1}=\frac{12EI}{L^3},\; M_1^{z_1}=\frac{6EI}{L^2},\; F_2^{z_1}=-\frac{12EI}{L^3},\; M_2^{z_1}=\frac{6EI}{L^2}$$

$$F_1^{\phi_1}=\frac{6EI}{L^2},\; M_1^{\phi_1}=\frac{4EI}{L},\; F_2^{\phi_1}=-\frac{6EI}{L^2},\; M_2^{\phi_1}=\frac{2EI}{L}$$

loading for column 3

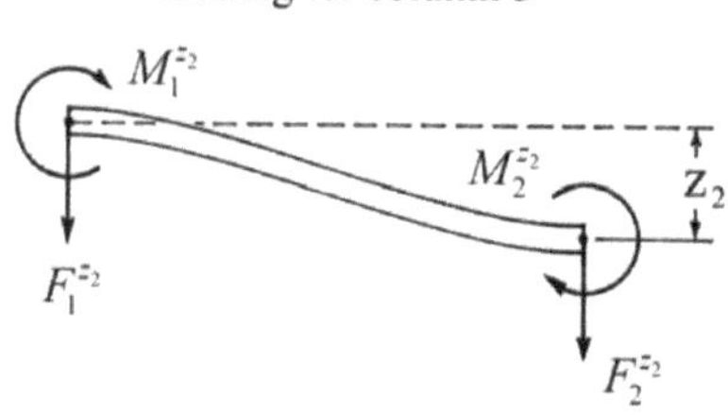

loading for column 4

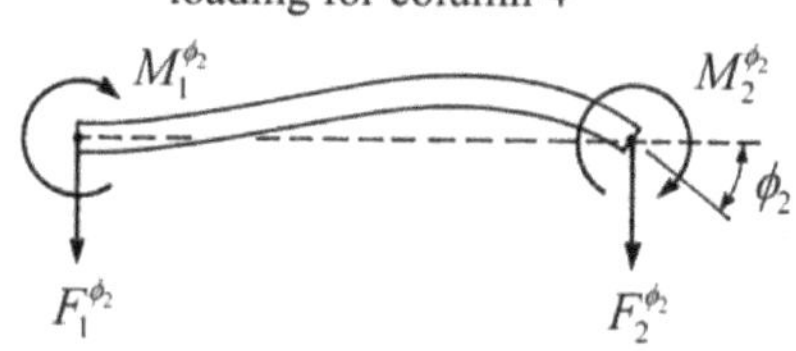

The loading for the third column is simply the negative of the first column:

$$F_1^{z_2}=-\frac{12EI}{L^3},\; M_1^{z_2}=-\frac{6EI}{L^2},\; F_2^{z_2}=\frac{12EI}{L^3},\; M_2^{z_2}=-\frac{6EI}{L^2}$$

And the loading for the fourth column is

$$F_1^{\phi_2} = \frac{6EI}{L^2},\ M_1^{\phi_2} = \frac{2EI}{L},\ F_2^{\phi_2} = -\frac{6EI}{L^2},\ M_2^{\phi_2} = \frac{4EI}{L}$$

Therefore, the complete stiffness influence coefficient matrix for the single uniform segment is

$$[K] = \begin{bmatrix} \frac{12EI}{L^3} & \frac{6EI}{L^2} & \frac{-12EI}{L^3} & \frac{6EI}{L^2} \\ \frac{6EI}{L^2} & \frac{4EI}{L} & \frac{-6EI}{L^2} & \frac{2EI}{L} \\ \frac{-12EI}{L^3} & \frac{-6EI}{L^2} & \frac{12EI}{L^3} & \frac{-6EI}{L^2} \\ \frac{6EI}{L^2} & \frac{2EI}{L} & \frac{-6EI}{L^2} & \frac{4EI}{L} \end{bmatrix}$$

5.4.3.3.1 **Aside**

It should be noted that this matrix is for a slender beam. A more exact representation would account for shear stresses, as described by the Timoshenko model (from J. S. Przemieniecki, *Theory of Matrix Structural Analysis*, McGraw-Hill, New York, 1968, p. 80):

$$[K] = \frac{EI}{L(1+\Phi)} \begin{bmatrix} 12/L^2 & 6/L & -12/L^2 & 6/L \\ 6/L & 4+\Phi & -6/L & 2-\Phi \\ -12/L^2 & -6/L & 12/L^2 & -6/L \\ 6/L & 2-\Phi & -6/L & 4+\Phi \end{bmatrix}$$

where

$$\Phi = \frac{12EI}{GA_{cs}L^2}$$

G is the shear modulus of elasticity, and A_{cs} is the cross-sectional area of the beam undergoing shear.

5.4.3.3.2 **End Aside**

The structural dynamics of a single-segment beam may now be studied.

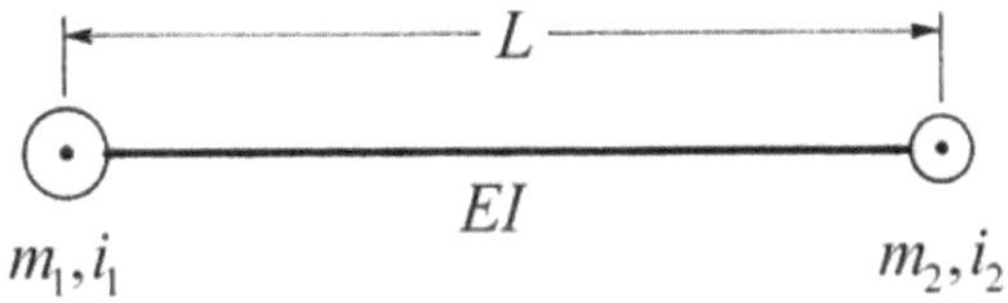

The force displacement equations may be written in terms of the stiffness matrix:

$$\begin{Bmatrix} F_1 \\ M_1 \\ F_2 \\ M_2 \end{Bmatrix} = [K]_{1,2} \begin{Bmatrix} z_1 \\ \phi_1 \\ z_2 \\ \phi_2 \end{Bmatrix} \tag{5.48}$$

where the forces and moments are due to inertial reactions:

$$F_1 = -m_1\ddot{z}_1, F_2 = -m_2\,\ddot{z}_2, M_1 = -i_1\ddot{\phi}_1, M_2 = -i_2\,\ddot{\phi}_2$$

In matrix form, these become

$$\begin{Bmatrix} F_1 \\ M_1 \\ F_2 \\ M_2 \end{Bmatrix} = -\begin{bmatrix} m_1 & 0 & 0 & 0 \\ 0 & i_1 & 0 & 0 \\ 0 & 0 & m_2 & 0 \\ 0 & 0 & 0 & i_2 \end{bmatrix} \begin{Bmatrix} \ddot{z}_1 \\ \ddot{\phi}_1 \\ \ddot{z}_2 \\ \ddot{\phi}_2 \end{Bmatrix} \equiv -[\mathcal{M}]_{1,2} \begin{Bmatrix} \ddot{z}_1 \\ \ddot{\phi}_1 \\ \ddot{z}_2 \\ \ddot{\phi}_2 \end{Bmatrix}$$

This, with Equation (5.48), gives the dynamic equations of the unforced single-segment beam:

$$[K]_{1,2} \begin{Bmatrix} z_1 \\ \phi_1 \\ z_2 \\ \phi_2 \end{Bmatrix} + [\mathcal{M}]_{1,2} \begin{Bmatrix} \ddot{z}_1 \\ \ddot{\phi}_1 \\ \ddot{z}_2 \\ \ddot{\phi}_2 \end{Bmatrix} = 0$$

This development may now be extended to a two-segment beam.

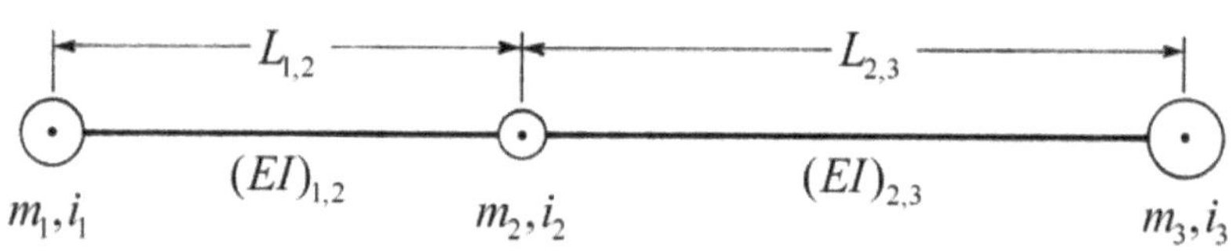

The stiffness matrix is 6 × 6, and the middle two columns may be found by the following argument. First of all, consider the loading for column 3:

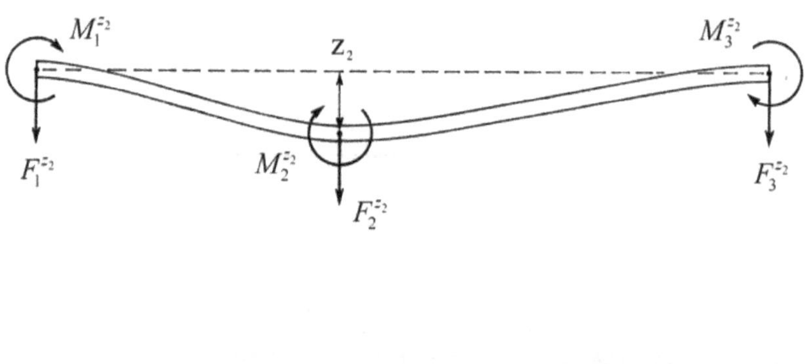

This may be conceptually divided at station 2 into two beams, loaded such that the conditions of tangency and displacement are met:

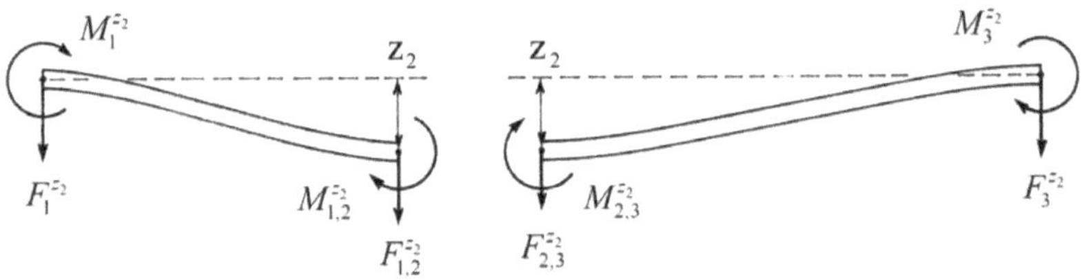

The forces and moments are obtained from previous work as

$$F_2^{z_2} = F_{1,2}^{z_2} + F_{2,3}^{z_2} = \left(\frac{12EI}{L^3}\right)_{1,2} + \left(\frac{12EI}{L^3}\right)_{2,3} = 12\left[\left(\frac{EI}{L^3}\right)_{1,2} + \left(\frac{EI}{L^3}\right)_{2,3}\right]$$

$$M_2^{z_2} = M_{1,2}^{z_2} + M_{2,3}^{z_2} = \left(\frac{-6EI}{L^2}\right)_{1,2} + \left(\frac{6EI}{L^2}\right)_{2,3} = -6\left[\left(\frac{EI}{L^2}\right)_{1,2} - \left(\frac{EI}{L^2}\right)_{2,3}\right]$$

Next, the loading for column 4 is shown as follows:

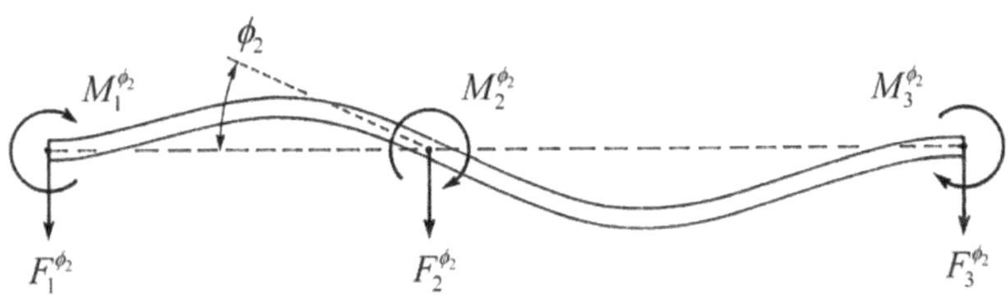

And as before, this may be conceptually divided:

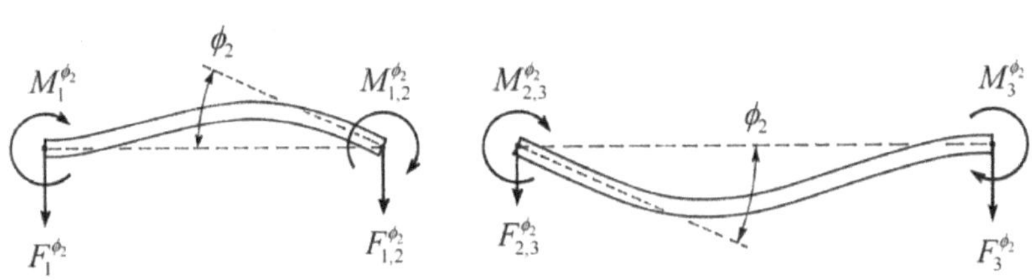

which gives the force and moment at station 2 as:

$$F_2^{\phi_2} = F_{1,2}^{\phi_2} + F_{2,3}^{\phi_2} = -\left(\frac{6EI}{L^2}\right)_{1,2} + \left(\frac{6EI}{L^2}\right)_{2,3} = 6\left[-\left(\frac{EI}{L^2}\right)_{1,2} + \left(\frac{EI}{L^2}\right)_{2,3}\right]$$

$$M_2^{\phi_2} = M_{1,2}^{\phi_2} + M_{2,3}^{\phi_2} = \left(\frac{4EI}{L}\right)_{1,2} + \left(\frac{4EI}{L}\right)_{2,3} = 4\left[\left(\frac{EI}{L}\right)_{1,2} + \left(\frac{EI}{L}\right)_{2,3}\right]$$

Therefore, the stiffness matrix is

$$[K]=\begin{bmatrix}
\left(\frac{12EI}{L^3}\right)_{1,2} & \left(\frac{6EI}{L^2}\right)_{1,2} & \left(\frac{-12EI}{L^3}\right)_{1,2} & \left(\frac{6EI}{L^2}\right)_{1,2} & 0 & 0 \\
\left(\frac{6EI}{L^2}\right)_{1,2} & \left(\frac{4EI}{L}\right)_{1,2} & \left(\frac{-6EI}{L^2}\right)_{1,2} & \left(\frac{2EI}{L}\right)_{1,2} & 0 & 0 \\
\left(\frac{-12EI}{L^3}\right)_{1,2} & \left(\frac{-6EI}{L^2}\right)_{1,2} & 12\left[\left(\frac{EI}{L^3}\right)_{1,2}+\left(\frac{EI}{L^3}\right)_{2,3}\right] & 6\left[-\left(\frac{EI}{L^2}\right)_{1,2}+\left(\frac{EI}{L^2}\right)_{2,3}\right] & \left(\frac{-12EI}{L^3}\right)_{2,3} & \left(\frac{-6EI}{L^2}\right)_{2,3} \\
\left(\frac{6EI}{L^2}\right)_{1,2} & \left(\frac{2EI}{L}\right)_{1,2} & 6\left[-\left(\frac{EI}{L^2}\right)_{1,2}+\left(\frac{EI}{L^2}\right)_{2,3}\right] & 4\left[\left(\frac{EI}{L}\right)_{1,2}+\left(\frac{EI}{L}\right)_{2,3}\right] & \left(\frac{-6EI}{L^2}\right)_{2,3} & \left(\frac{2EI}{L}\right)_{2,3} \\
0 & 0 & \left(\frac{-12EI}{L^3}\right)_{2,3} & \left(\frac{-6EI}{L^2}\right)_{2,3} & \left(\frac{12EI}{L^3}\right)_{2,3} & \left(\frac{-6EI}{L^2}\right)_{2,3} \\
0 & 0 & \left(\frac{6EI}{L^2}\right)_{2,3} & \left(\frac{2EI}{L}\right)_{2,3} & \left(\frac{-6EI}{L^2}\right)_{2,3} & \left(\frac{4EI}{L}\right)_{2,3}
\end{bmatrix}$$

Note that this matrix has the form of two overlapping single-element matrices:

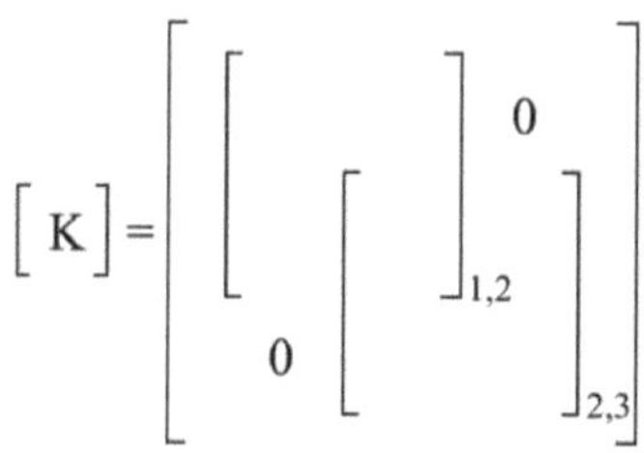

This may be generalized to obtain the stiffness matrix for a beam of n stations:

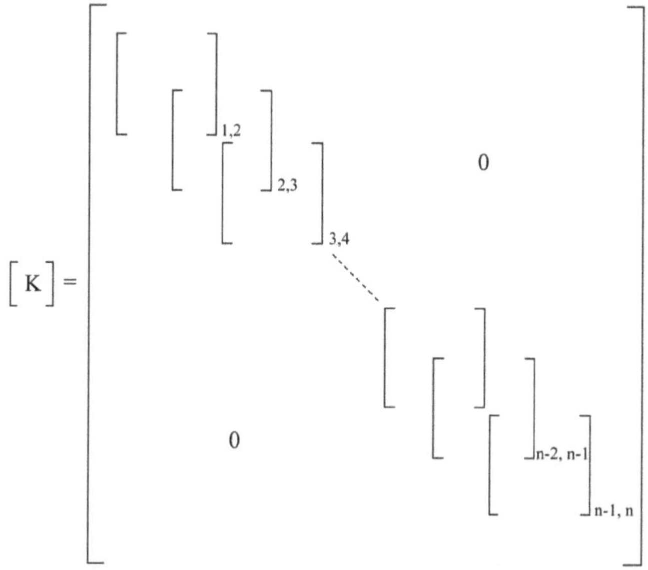

Note that this construction of the stiffness matrix from discrete elements is a fundamental technique in finite element theory, called the direct stiffness method. A detailed explanation of its general application may be found in *A First Course in the Finite Element Method Using Algor* by Daryl Logan, Brooks/Cole 2001.

Now, the corresponding mass matrix is

$$[\mathcal{M}]=\begin{bmatrix} m_1 & & & & & & & & \\ & i_1 & & & & & & & \\ & & m_2 & & & & & 0 & \\ & & & i_2 & & & & & \\ & & & & \cdot & & & & \\ & & & & & \cdot & & & \\ & & 0 & & & & \cdot & & \\ & & & & & & & m_n & \\ & & & & & & & & i_n \end{bmatrix}$$

Therefore, the unforced dynamic equations are

$$[K]\begin{Bmatrix} z_1 \\ \phi_1 \\ z_2 \\ \phi_2 \\ \cdot \\ \cdot \\ \cdot \\ z_n \\ \phi_n \end{Bmatrix}+[\mathcal{M}]\begin{Bmatrix} \ddot{z}_1 \\ \ddot{\phi}_1 \\ \ddot{z}_2 \\ \ddot{\phi}_2 \\ \cdot \\ \cdot \\ \cdot \\ \ddot{z}_n \\ \ddot{\phi}_n \end{Bmatrix}=0$$

One may rearrange the $[K]$ and $[\mathcal{M}]$ matrices to give

$$\begin{bmatrix} [K^{zz}] & [K^{z\phi}] \\ [K^{\phi z}] & [K^{\phi\phi}] \end{bmatrix}\begin{Bmatrix} \{z\} \\ \{\phi\} \end{Bmatrix}=-\begin{bmatrix} [M] & 0 \\ 0 & [i] \end{bmatrix}\begin{Bmatrix} \{\ddot{z}\} \\ \{\ddot{\phi}\} \end{Bmatrix}$$

Assume that all the moment-of-inertia terms are negligible, so that $[i]\approx 0$. The equations then become

$$[K^{zz}]\{z\}+[K^{z\phi}]\{\phi\}=-[M]\{\ddot{z}\}$$

$$[K^{\phi z}]\{z\}+[K^{\phi\phi}]\{\phi\}=0$$

And these combine to give

$$[M]\{\ddot{z}\}+\left[\left[K^{zz}\right]-\left[K^{z\phi}\right]\left[K^{\phi\phi}\right]^{-1}\left[K^{\phi z}\right]\right]\{z\}\equiv[M]\{\ddot{z}\}+\left[\bar{K}\right]\{z\}=0$$

If the beam is free-free, then $\left[\bar{K}\right]$ is singular. However, if the beam is supported, then $\left[\bar{K}\right]$ is non-singular and

$$\left[C^{zz}\right]=\left[\bar{K}\right]^{-1}$$

Finally, the dynamic equations are

$$[M]\{\ddot{z}\}+\left[\bar{K}\right]\{z\}=0$$

By assuming the solution $z=Z\sin\omega t$, one obtains an eigenvalue problem for the vibration frequencies and corresponding modes:

$$\left[\bar{K}-\omega^2 M\right]\{Z\}=0 \tag{5.49}$$

5.4.3.3.3 *Numerical Example*

Consider a uniform free-free beam with the following properties:

$$L_{total}=12.192\ m;\ m(y)=35.7087\ kg/m;\ EI(y)=9.7523\times10^6\ N-m^2$$

This will be modelled with a four-segment beam:

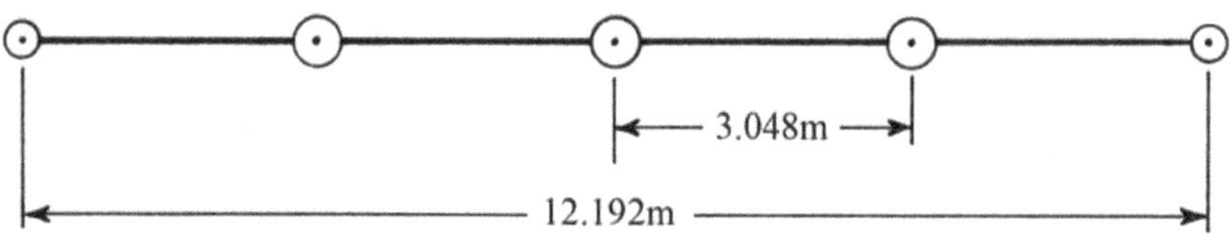

The five lumped masses are determined from

$$m_1=m\,L_{1,2}/2=54.42\ kg,\ m_5=m\,L_{4,5}/2=54.42\ kg$$

$$m_i=m\left(L_{i-1,i}+L_{i,i+1}\right)/2=108.84\ kg,\ i=2,3,4$$

and

$$(EI)_{i,i+1}=9.7523\times10^6\ N-m^2,\ L_{i,i+1}=3.048\ m,\ i=1\ to\ 4$$

The resulting $\left[K\right]$ matrix is

$$[K]=\begin{bmatrix} 4.133 & 6.298 & -4133 & 6.298 & 0 & 0 & 0 & 0 & 0 & 0 \\ 6.298 & 12.798 & -6.298 & 6.399 & 0 & 0 & 0 & 0 & 0 & 0 \\ -4.133 & -6.298 & 8.266 & 0 & -4.133 & 6.298 & 0 & 0 & 0 & 0 \\ 6.298 & 6.399 & 0 & 25.597 & -6.298 & 6.399 & 0 & 0 & 0 & 0 \\ 0 & 0 & -4.133 & -6.298 & 8.266 & 0 & -4.133 & 6.298 & 0 & 0 \\ 0 & 0 & 6.298 & 6.399 & 0 & 25.597 & -6.298 & 6.399 & 0 & 0 \\ 0 & 0 & 0 & 0 & -4.133 & -6.298 & 8.266 & 0 & -4.133 & 6.298 \\ 0 & 0 & 0 & 0 & 6.298 & 6.399 & 0 & 25.597 & -6.298 & 6.399 \\ 0 & 0 & 0 & 0 & 0 & 0 & 4.133 & -6.298 & 4.133 & -6.298 \\ 0 & 0 & 0 & 0 & 0 & 0 & 6.298 & 6.399 & -6.298 & 12.798 \end{bmatrix} x10^6$$

and the partitioned matrices are

$$\left[K^{zz}\right]=\begin{bmatrix} 4.1328 & -4.1328 & 0 & 0 & 0 \\ -4.1328 & 8.2656 & -4.1328 & 0 & 0 \\ 0 & -4.1328 & 8.2656 & -4.1328 & 0 \\ 0 & 0 & -4.1328 & 8.2656 & -4.1328 \\ 0 & 0 & 0 & -4.1328 & 4.1328 \end{bmatrix}\times 10^6$$

$$\left[K^{z\phi}\right]=\begin{bmatrix} 6.2984 & 6.2984 & 0 & 0 & 0 \\ -6.2984 & 0 & 6.2984 & 0 & 0 \\ 0 & -6.2984 & 0 & 6.2984 & 0 \\ 0 & 0 & -6.2984 & 0 & 6.2984 \\ 0 & 0 & 0 & -6.2984 & -6.2984 \end{bmatrix}\times 10^6$$

$$\left[K^{\phi z}\right]=\begin{bmatrix} 6.2984 & -6.2984 & 0 & 0 & 0 \\ 6.2984 & 0 & -6.2984 & 0 & 0 \\ 0 & 6.2984 & 0 & -6.2984 & 0 \\ 0 & 0 & 6.2984 & 0 & -6.2984 \\ 0 & 0 & 0 & 6.2984 & -6.2984 \end{bmatrix}\times 10^6$$

$$\left[K^{\phi\phi}\right]=\begin{bmatrix} 1.2798 & 0.6399 & 0 & 0 & 0 \\ 0.6399 & 2.5597 & 0.6399 & 0 & 0 \\ 0 & 0.6399 & 2.5597 & 0.6399 & 0 \\ 0 & 0 & 0.6399 & 2.5597 & 0.6399 \\ 0 & 0 & 0 & 0.6399 & 1.2798 \end{bmatrix}\times 10^7$$

The resulting $\left[\bar{K}\right]$ matrix is

$$\left[\bar{K}\right]=\begin{bmatrix} 0.5535 & -1.2546 & 0.8856 & -0.2214 & 0.0369 \\ -1.2546 & 3.3948 & -3.2472 & 1.3284 & -0.2214 \\ 0.8856 & -3.2472 & 4.7232 & -3.2472 & 0.8856 \\ -0.2214 & 1.3284 & -3.2472 & 3.3948 & -1.2546 \\ 0.0369 & -0.2214 & 0.8856 & -1.2546 & 0.5535 \end{bmatrix}\times 10^6$$

From this one obtains the five modal solutions, for which the fundamental bending mode is

$$\{Z\}_1=\begin{Bmatrix} -1.0000 \\ 0.1650 \\ 0.6701 \\ 0.1650 \\ -1.0000 \end{Bmatrix},\ \omega_1=66.48\ rad\,/\,\mathrm{sec}$$

For the free-free beam in Section 5.2, the exact solution is

$$\omega_1=\frac{22.37}{L_{total}^2}\left(\frac{EI}{m}\right)^{1/2}=78.65\ rad\,/\,\mathrm{sec}$$

The 4-segment beam is, of course, a crude approximation to the continuous beam; but it serves as an academic example. If a 20-segment beam is chosen,

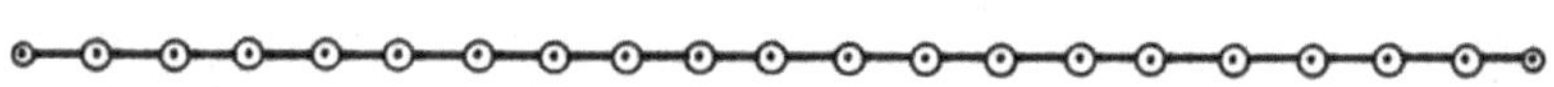

then the fundamental frequency is 78.1 *rad/sec*. This is clearly converging to the exact solution.

The obvious must be stated: the small circles representing the concentrated masses are symbolic only. The massless elastic segments extend from the center of one circle to the next.

For finite-wing flutter, discussed in Section 6.4, it is commonly assumed that the wing may be modelled by a cantilever beam because the inertia of the center body of the airplane suppresses its vertical acceleration and causes the free-free beam to act like it was clamped in the middle. It is possible to quantify this with the previous 20-segment beam example by adding a center mass to the middle concentrated mass:

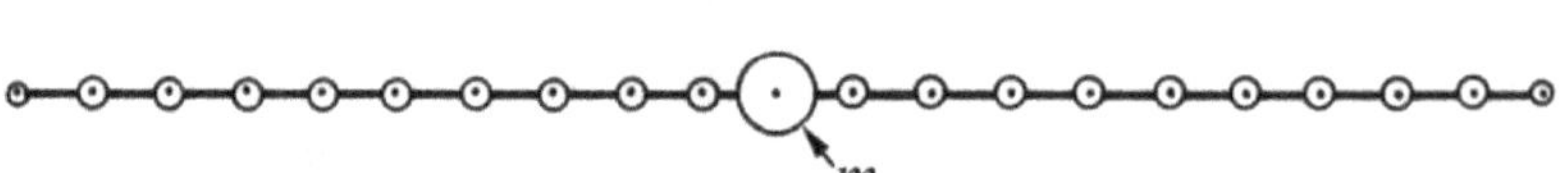

$$\left(m_{11}\right)_{new}=\left(m_{11}\right)_{previous}+m_{centre}$$

Also, note that the total mass of the wing (minus the center mass) is

$$m_{wing} = m \times L_{total} = 435.3605\ kg$$

Now, if the first mode (fundamental) frequency is plotted versus the mass ratio, defined as

$$Mass\ Ratio = m_{centre} / m_{wing}$$

one obtains the following graph:

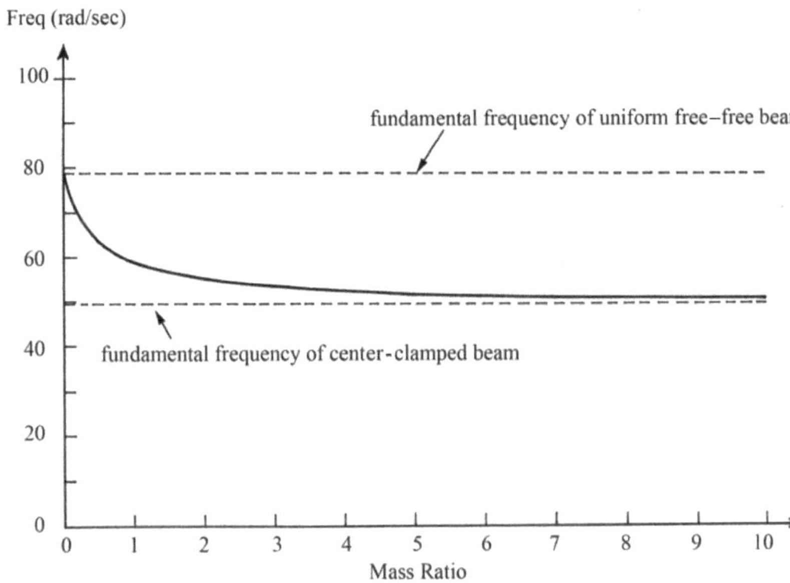

This may be related to the ratio of wing mass over the total airplane mass by

$$\frac{m_{wing}}{m_{total}} = \frac{m_{wing}}{\left(m_{wing} + m_{centre}\right)} = \frac{1}{\left(1 + Ratio\right)} \rightarrow Ratio = \frac{1}{m_{wing} / m_{total}} - 1$$

For most conventional airplanes with a central fuselage (especially if the engine and undercarriage are attached to the fuselage), Torenbeek (*Synthesis of Subsonic Airplane Design*, Delft University Press, 1982, pp. 279–280) shows that

$$m_{wing} / m_{total} \approx 0.10$$

which gives

$$Ratio \approx 9$$

It is seen from the previous graph that this range is close to that for the clamped beam case; thus, the cantilever assumption is reasonably justified. It is important to note, however, that not all airplanes fall within this range. Span-loading aircraft, where the engines and stores are distributed along the wing, may behave much closer to the unrestrained free-free beam. An excellent example of this is the solar-powered flying-wing design from AeroVironment Incorporated:

NASA Dryden Flight Research Center Photo Collection
http://www.dfrc.nasa.gov/gallery/photo/index.html
NASA Photo: ED01-0230-1 Date: August 13, 2001 Photo by: Nick Galante
NASA's Helios Prototype aircraft taking off from the Pacific Missile Range Facility, Kauai, Hawaii, for the record flight.

In truth, this is a remarkable physical manifestation of the segmented-beam model.

5.4.3.4 Assumed-Mode Method for the Lumped-Parameter Beam

Consider the transformation

$$\{z\}=[Q]\{q\} \tag{5.50}$$

This may be used to change the problem from finding the $\{z\}$ displacements at specific stations along the beam to that of finding the amplitudes, $\{q\}$, of assumed modes. First of all, consider the kinetic energy:

$$T=\frac{1}{2}\dot{z}[M]\{\dot{z}\}$$

With Equation (5.50), this becomes

$$T=\frac{1}{2}\dot{q}[\tilde{M}]\{\dot{q}\}$$

where

$$\left[\tilde{M}\right]=\left[Q\right]'\left[M\right]\left[Q\right] \tag{5.51}$$

Likewise, one has the strain energy:

$$V=\frac{1}{2}z\left[\bar{K}\right]\{z\}$$

which becomes

$$V=\frac{1}{2}q\left[\tilde{K}\right]\{q\} \tag{5.52}$$

where

$$\left[\tilde{K}\right]=\left[Q\right]'\left[\bar{K}\right]\left[Q\right]$$

Now, consider a matrix of forces applied in the z direction at stations j:

$$F_j=\{F\},\ j=1,2,3,\ldots\ldots\ldots,\ n$$

Also, from Equation (5.50), the $\dot{z}$ velocities at stations j are given by

$$\{\dot{z}\}=\left[Q\right]\{\dot{q}\}\rightarrow \dot{z}_j=\sum_{i=1}^{n}Q_{ji}\,\dot{q}_i$$

Therefore, the generalized forces are

$$\tilde{Q}_i=\sum_{j=1}^{n}Q_{ji}\,F_j=\left[Q\right]'\{F\}\equiv\{\tilde{Q}\} \tag{5.53}$$

and Equations (5.51), (5.52), and (5.53) combine in Lagrange's energy equation to give the dynamic equations

$$\left[\tilde{M}\right]\{\ddot{q}\}+\left[\tilde{K}\right]\{q\}=\{\tilde{Q}\} \tag{5.54}$$

Note that $\left[Q\right]$ is composed of columns that describe the assumed modes:

$$\left[Q\right]=\left[\{Z\}_1\,\{Z\}_2\cdots\cdots\{Z\}_n\right]$$

where
1. The $\{Z\}_i$ column vectors satisfy the end conditions
2. The vectors are independent (none are a linear combination of the others)

5.4.3.5 Normal Modes

Normal modes are defined to be the $\{Z\}$ eigenvectors of the *unforced* beam under consideration. These may be used as assumed modes for the forced-beam problem

(Equation (5.54)) because their orthogonality gives $\left[\tilde{M}\right]$ and $\left[\tilde{K}\right]$ matrices that are diagonal.

Proof: Consider the solution of two modes for the unforced lumped-parameter beam (Equation 5.49):

$$\left[\bar{K}\right]\{Z\}_i = \omega_i^2 [M]\{Z\}_i \tag{5.55}$$

$$\left[\bar{K}\right]\{Z\}_j = \omega_j^2 [M]\{Z\}_j \tag{5.56}$$

Now, $Z_j \times (5.55)$ and $Z_i \times (5.56)$ give

$$Z_j\left[\bar{K}\right]\{Z\}_i = \omega_i^2 Z_j [M]\{Z\}_i \tag{5.57}$$

$$Z_i\left[\bar{K}\right]\{Z\}_j = \omega_j^2 Z_i [M]\{Z\}_j \tag{5.58}$$

Note that

$$Z_j\left[\bar{K}\right]\{Z\}_i = Z_i\left[\bar{K}\right]\{Z\}_j$$

so, (5.57) minus (5.58) gives

$$\left(\omega_i^2 - \omega_j^2\right) Z_j [M]\{Z\}_i = 0$$

Therefore, for distinct frequencies, one has

$$Z_j [M]\{Z\}_i = \tilde{M}_{ij} = 0 \; for \, i \neq j$$

This means that $\left[\tilde{M}\right]$ is diagonal:

$$[Q]'[M][Q] = \begin{bmatrix} \tilde{M}_{11} & & 0 \\ & \tilde{M}_{22} & \\ & & \cdot \\ 0 & & \tilde{M}_{nn} \end{bmatrix}$$

Further, Equation (5.55) gives that

$$Z_j\left[\bar{K}\right]\{Z\}_i = \omega_i^2 Z_j [M]\{Z\}_i = \omega_i^2 \left[\tilde{M}_{ij}\right]$$

so

$$Z_j\left[\bar{K}\right]\{Z\}_i = \tilde{K}_{ij} = 0 \; for \, i \neq j$$

which means that $\left[\tilde{K}\right]$ is diagonal:

$$[Q]'\left[\bar{K}\right][Q]=\begin{bmatrix}\tilde{K}_{11} & & & 0\\ & \tilde{K}_{22} & & \\ & & \cdot & \\ 0 & & & \tilde{K}_{nn}\end{bmatrix}$$

This completes the proof.

Now, consider the forced-beam problem again (Equation (5.54)). If $\{F\}$ is a function of $\{q\}$, $\{\dot{q}\}$, $\{\ddot{q}\}$, etc., then coupling among the equations occurs on the RHS only:

$$\begin{bmatrix}\tilde{M}_{11} & & & 0\\ & \tilde{M}_{22} & & \\ & & \cdot & \\ 0 & & & \tilde{M}_{nn}\end{bmatrix}\begin{Bmatrix}\ddot{q}_1\\ \ddot{q}_2\\ \cdot\\ \ddot{q}_n\end{Bmatrix}+\begin{bmatrix}\tilde{K}_{11} & & & 0\\ & \tilde{K}_{22} & & \\ & & \cdot & \\ 0 & & & \tilde{K}_{nn}\end{bmatrix}\begin{Bmatrix}q_1\\ q_2\\ \cdot\\ q_n\end{Bmatrix}=\begin{Bmatrix}\tilde{Q}_1\\ \tilde{Q}_2\\ \cdot\\ \tilde{Q}_n\end{Bmatrix} \tag{5.59}$$

However, if $\{F\}$ is a function of time only, then Equations (5.59) are functions of time *only* and may be solved for individually:

$$\begin{aligned}\tilde{M}_{11}\ddot{q}_1+\tilde{K}_{11}q_1&=\tilde{Q}_1\\ \tilde{M}_{22}\ddot{q}_2+\tilde{K}_{22}q_2&=\tilde{Q}_2\\ \cdot\qquad\quad \cdot\qquad &\quad\cdot\\ \tilde{M}_{nn}\ddot{q}_n+\tilde{K}_{nn}q_n&=\tilde{Q}_n\end{aligned}$$

This shows the usefulness of the unforced solutions for solving the forced-beam problem.

5.4.3.6 Normal Mode Example

This example consists of three identical masses, m, on a stretched string. Note that this is *not* a lumped-parameter beam. However, it does a very good job of elucidating normal modes and normal coordinates.

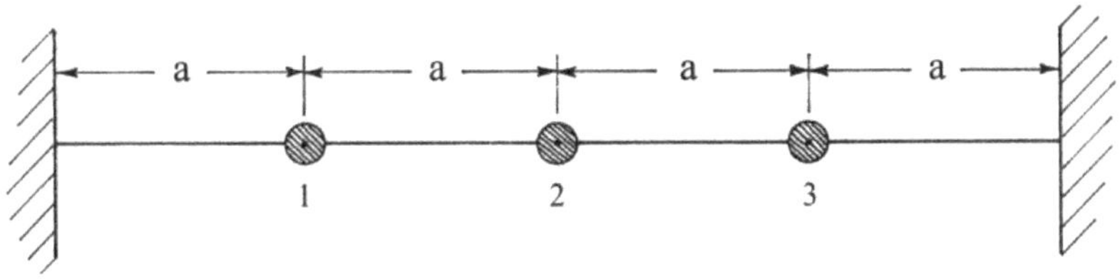

The dynamic equations are

$$m\ddot{z}_1+S/a\left(2z_1-z_2\right)=0$$

$$m\ddot{z}_2 + S/a\left(-z_1 + 2z_2 - z_3\right) = 0$$

$$m\ddot{z}_3 + S/a\left(-z_2 + 2z_3\right) = 0$$

where S is the string tension and z_1, z_2, and z_3 are the vertical displacements at stations 1, 2, and 3.

The solutions give the three frequencies:

$$\omega_1^2 = \frac{\left(2-\sqrt{2}\right)S}{am},\ \omega_2^2 = \frac{2S}{am},\ \omega_3^2 = \frac{\left(2+\sqrt{2}\right)S}{am}$$

And the three modal vectors are (normalized with respect to z_3)

$$\{Z_1\} = \begin{Bmatrix} 1 \\ \sqrt{2} \\ 1 \end{Bmatrix}, \{Z_2\} = \begin{Bmatrix} -1 \\ 0 \\ 1 \end{Bmatrix}, \{Z_3\} = \begin{Bmatrix} 1 \\ -\sqrt{2} \\ 1 \end{Bmatrix}$$

Now, new coordinates may be defined by

$$\{z\} = [Q]\{q\}$$

where

$$[Q] = \left[\{Z\}_1\ \{Z\}_2\ \{Z\}_3\right]$$

so that

$$\begin{Bmatrix} z_1 \\ z_2 \\ z_3 \end{Bmatrix} = \begin{bmatrix} 1 & -1 & 1 \\ \sqrt{2} & 0 & -\sqrt{2} \\ 1 & 1 & 1 \end{bmatrix} \begin{Bmatrix} q_1 \\ q_2 \\ q_3 \end{Bmatrix}$$

The inverse of this gives

$$q_1 = z_1 + \sqrt{2}z_2 + z_3$$

$$q_2 = -z_1 + z_3$$

$$q_3 = z_1 - \sqrt{2}z_2 + z_3$$

Thus, the normal coordinates, q_i, are expressed in terms of the original coordinates z_i. A physical interpretation may be obtained from the system's modes of motion:

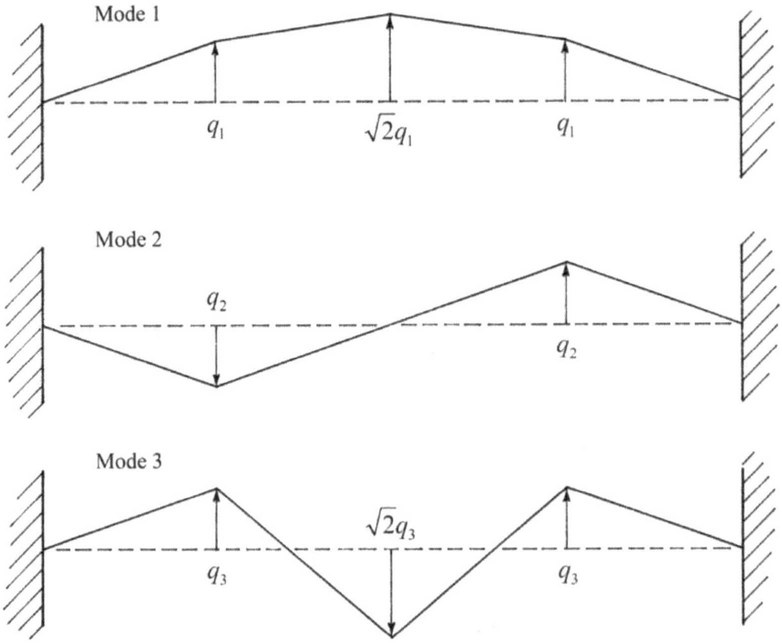

Note that the shape of each mode is determined by the eigenvectors, $\{Z\}_i$, and the scale of each mode is proportional to the magnitude of q_i. Therefore, a normal coordinate simply describes the *magnitude* of a particular mode of the system.

6 Classical Flutter

6.1 INTRODUCTION

Classical flutter consists of the coupling of two or more degrees of freedom. Also, the flow is assumed to be fully attached. By contrast, non-classical flutter consists primarily of the excitation of structures from the pressure variations due to separated flow. Examples of this are stall flutter, transmission line (and tower) gallop, and tab buzz.

Now, consider the vibration of an aircraft structure when disturbed by a gust or pilot input.

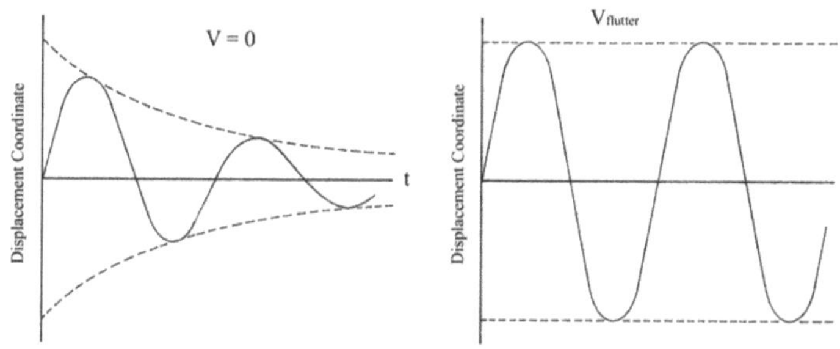

When the airspeed is zero (aircraft at rest), the vibrations will damp out with time due to mechanical or material damping. However, the *flutter velocity* is defined to be that value at which the input energy from the airspeed exactly balances the damping energy. Therefore, the vibratory condition is sustained. At higher speeds, the amplitude may grow, causing structural failure.

The level of flutter is measured by a damping parameter, g, which is usually plotted against velocity, V. Typical $g - V$ plots are shown as follows:

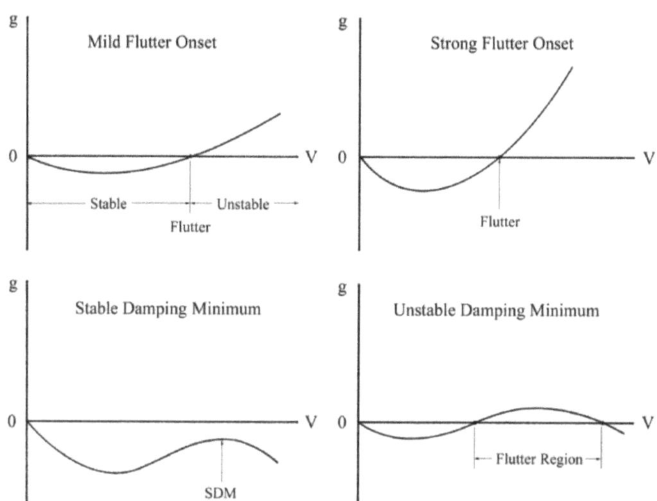

DOI: 10.1201/9781032709093-6

The way to interpret these is to consider g to be that value required to achieve flutter at any given V. For example, if g is negative for a certain V, then one would have to apply negative structural damping to make the system flutter. This, of course, is a stable situation because negative structural damping is unlikely. By contrast, if g is positive, then positive structural damping is required to stabilize the system. These plots show the degree of damping required.

6.2 REPRESENTATIVE SECTION

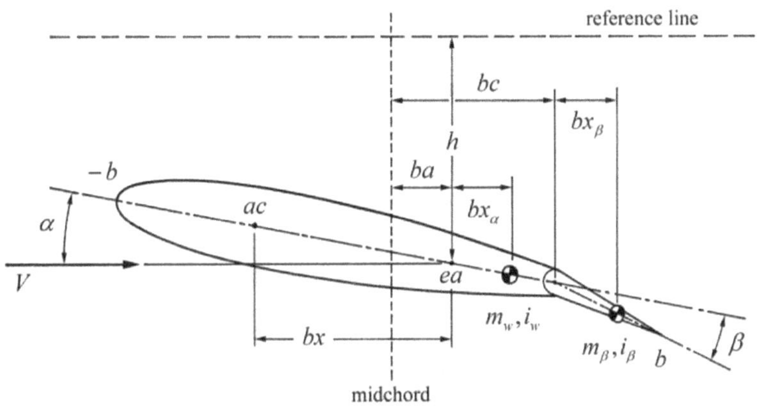

The motion coordinates are plunging, h; pitching, α; and control surface deflection, β. Note that α is strictly a geometrical measure of the main chord's angle with respect to the horizontal (this is in contrast to the definition of α used in flight dynamics).

6.2.1 INERTIAL TERMS

The inertial-reaction force and moments on the section are given by

$$\begin{Bmatrix} (F_h)_{inertial} \\ (M_\alpha)_{inertial} \\ (M_\beta)_{inertial} \end{Bmatrix} = \begin{bmatrix} M & S_\alpha & S_\beta \\ S_\alpha & I_\alpha & P_{\alpha\beta} \\ S_\beta & P_{\alpha\beta} & I_\beta \end{bmatrix} \begin{Bmatrix} \ddot{h} \\ \ddot{\alpha} \\ \ddot{\beta} \end{Bmatrix}$$

where F_h is the vertical force at ea, M_α is the moment about ea, and M_β is the moment about the control-surface hinge.

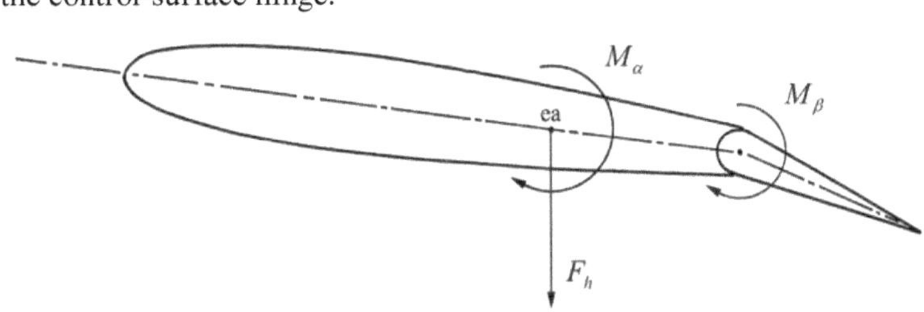

These may be found by a force and moment balance of the component inertial-reaction forces and moments, or energy methods may be used. That is, the system's kinetic energy may be expressed as a function of the motion variables, and then Lagrange's equation may be used to find the net reaction force and moments. When this is done, one obtains that

$$M = m_w + m_\beta$$
$$S_\alpha = m_w b x_\alpha + m_\beta \left[b(c-a) + b x_\beta \right]$$
$$S_\beta = m_\beta b x_\beta$$
$$I_\alpha = i_w + i_\beta + m_w b^2 x_\alpha^2 + m_\beta \left[b(c-a) + b x_\beta \right]^2$$
$$P_{\alpha\beta} = i_\beta + S_\beta \left[b(c-a) + b x_\beta \right]$$
$$I_\beta = i_\beta + m_\beta b^2 x_\beta^2$$

6.2.2 Elastic Terms

The elasticity of the section is modeled by the system of springs shown here:

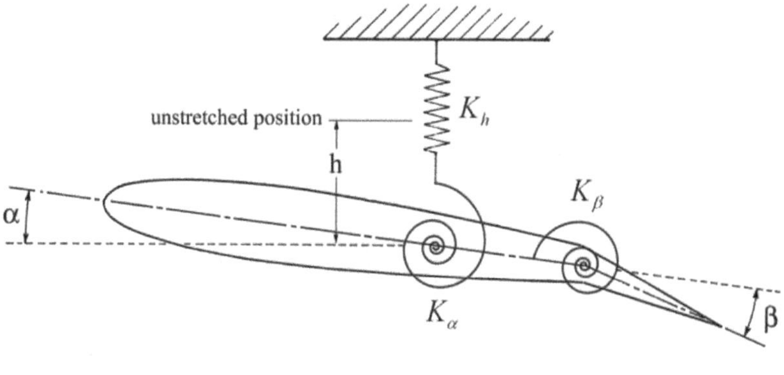

From this it can be readily seen that the elastic reaction forces are given by

$$\begin{Bmatrix} (F_h)_{elastic} \\ (M_\alpha)_{elastic} \\ (M_\beta)_{elastic} \end{Bmatrix} = \begin{bmatrix} K_h & 0 & 0 \\ 0 & K_\alpha & 0 \\ 0 & 0 & K_\beta \end{bmatrix} \begin{Bmatrix} h \\ \alpha \\ \beta \end{Bmatrix}$$

6.2.3 Structural Damping Terms

Structural damping arises from:

1. Friction at the connections, such as riveted joints (in general, there is some "working" between the connected pieces, with consequent friction damping).
2. Dissipation within the material itself ("material damping").

From experiments, it has been found that such damping is:

1. Proportional to displacement
2. Independent of frequency
3. In phase with velocity

Mathematically, this may be expressed by replacing the stiffness terms, K, with $K(1+ig)$, where g is the structural damping parameter:

$$K_h \to K_h(1+ig_h),\ K_\alpha \to K_\alpha(1+ig_\alpha),\ K_\beta \to K_\beta(1+ig_\beta)$$

6.2.3.1 Example

Consider a simple harmonic system:

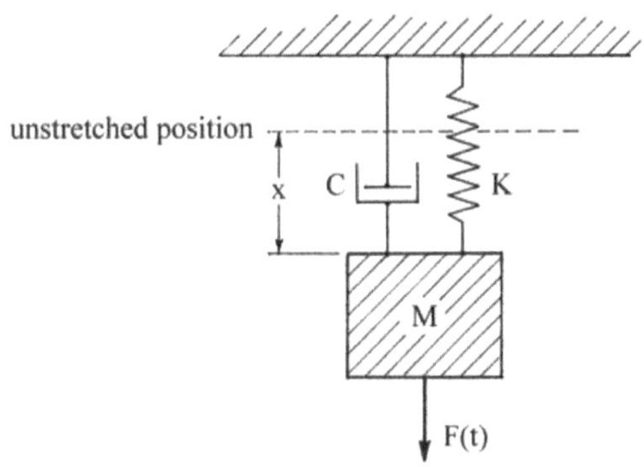

$$M\ddot{x} + C\dot{x} + Kx = F(t)$$

For the structural damping model, this becomes

$$M\ddot{x} + K(1+ig)x = F(t) \tag{6.1}$$

Now, if the system is oscillating with simple harmonic motion *only* (this would be the case if $C, g \ll 1$ *or* flutter), one has that

$$x = e^{i\omega t} \to \dot{x} = i\omega e^{i\omega t} = i\omega x$$

so that

$$iKgx = \frac{Kg}{\omega}\dot{x}$$

Therefore, Equation (6.1) becomes

$$M\ddot{x} + \frac{Kg}{\omega}\dot{x} + Kx = F(t)$$

which shows an equivalence between viscous and structural damping *for this special case*:

$$C = \frac{Kg}{\omega} \text{ for simple harmonic motion} \tag{6.2}$$

6.2.3.2 Aside

Consider a very lightly-damped unforced system:

$$M\ddot{x} + C\dot{x} + Kx = 0, C \ll 1$$

The motion is approximately given by

$$x = e^{i\omega t}$$

so the equation becomes

$$-M\omega^2 + i\omega C + K = 0 \rightarrow \omega^2 - i\frac{\omega C}{M} - \frac{K}{M} = 0 \tag{6.3}$$

Compare this with the form

$$\lambda^2 + 2\zeta\omega_n\lambda + \omega_n^2 = 0$$

For very small damping, one has that

$$\lambda \approx i\omega \approx i\omega_n \rightarrow \omega^2 - i2\zeta\omega^2 - \omega^2 = 0 \tag{6.4}$$

Upon comparing Equations (6.3) and (6.4), one obtains

$$\zeta = \frac{C}{2\omega M},\ \omega^2 = \frac{K}{M}$$

which gives that

$$\zeta = \frac{C\omega}{2K}$$

Upon comparison with Equation (6.2), one obtains that

$$g = 2\zeta \equiv 2 \times (\textit{Critical Damping Ratio})$$

This provides a means for determining g from natural-vibration tests on example structures. From experiment, g varies from 0.02 to 0.08 (riveted structure). Note that g is dimensionless.

6.2.3.3 End Aside

From Equation (6.2), one may express the damping forces as

$$\begin{Bmatrix} (F_h)_{damping} \\ (M_\alpha)_{damping} \\ (M_\beta)_{damping} \end{Bmatrix} = \begin{bmatrix} K_h g_h / \omega & 0 & 0 \\ 0 & K_\alpha g_\alpha / \omega & 0 \\ 0 & 0 & K_\beta g_\beta / \omega \end{bmatrix} \begin{Bmatrix} \dot{h} \\ \dot{\alpha} \\ \dot{\beta} \end{Bmatrix}$$

Again, note that this is only true for simple harmonic motion in h, α, and β.

6.2.4 Aerodynamic Forcing Functions

The aerodynamic force and moments are shown as follows. The prime denotes that these are perturbations from the reference equilibrium values.

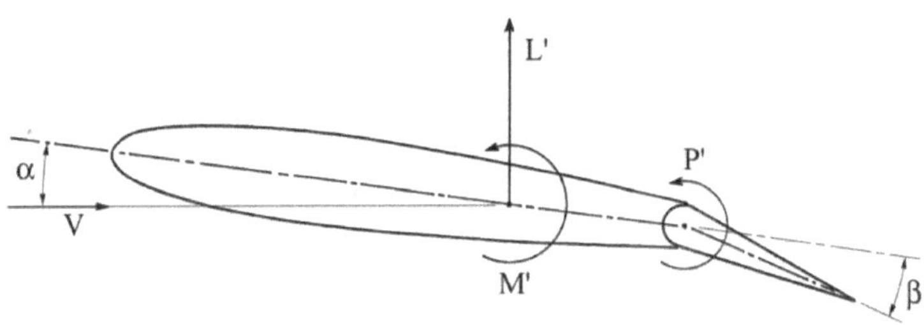

The typical frequencies of oscillation for flutter are usually such that quasi-steady aerodynamic models are inadequate. The parameter that determines this is the "reduced frequency," k, which is defined as

$$k = \frac{\omega c}{2V} = \frac{\omega b}{V}$$

where c is the chord of the airfoil ($c = 2b$). Another way to look at this is with a parameter from flapping-wing theory, the "advance ratio," λ, given by

$$\lambda = \frac{V}{cf} = \frac{\pi}{k}$$

where f is the oscillation frequency in cycles per second. λ is the distance travelled in chord lengths per flapping cycle, and the significance of this is that it provides a measure of the influence of the shed wake on the airfoil's pressure distribution. For subsonic flight, the following figure shows a shed and bound vortex on an airfoil:

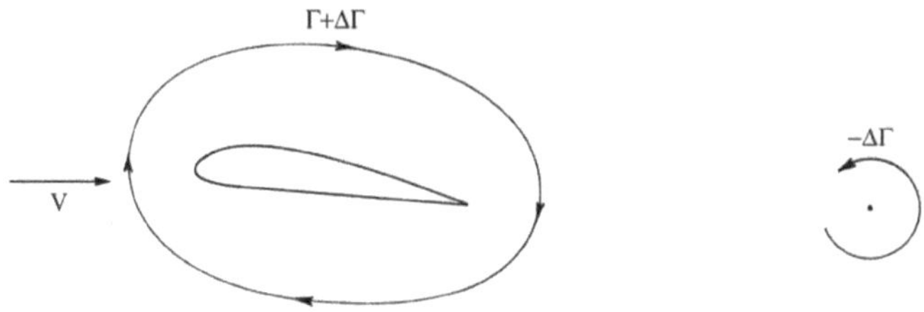

For every change of angle of attack, there is a corresponding change in the bound vorticity, Γ. However, from conservation of vorticity, this change, $\Delta\Gamma$, is accompanied by a shed vortex of strength $-\Delta\Gamma$. For oscillatory motion, Γ changes continuously and leaves a trailing vortex wake. This wake, in turn, modifies the forces experienced by the airfoil from those for steady conditions. A way to look at this is to recall that the airfoil imposes boundary conditions on the flow and, for the steady-state case, the bound vorticity is determined by the requirement to satisfy the surface tangential-flow conditions when the incident flow is the free-stream velocity, V. However, when a shed vortex wake imposes an additional flow state on the airfoil, the bound vorticity has to adjust in order that the tangential-flow condition remains satisfied. This modifies the pressure distribution from that for the steady-state case. For $\beta = 0$ and incompressible flow, the lift expression is

$$L' = \rho b^2 \pi \left(V\dot{\alpha} + \ddot{h} - ba\ddot{\alpha} \right) + 2\pi\rho VbC(k)\left[V\alpha + \dot{h} + b(1/2 - a)\dot{\alpha} \right] \tag{6.5}$$

The first term is a solution that satisfies the boundary conditions without circulation (Kutta condition is not satisfied). Note that

$$w_n \equiv V\alpha + \dot{h} - ba\dot{\alpha}$$

where w_n is the normal component of the relative wind vector to the midchord. Therefore, $\dot{w}_n$ is the time rate of change of the relative velocity, w_n, to the chord ("acceleration"), so the coefficient of $\dot{w}_n$ is called the apparent (or virtual) mass:

$$m_{app} = \rho b^2 \pi$$

A way to interpret this is to observe that $\rho b^2 \pi$ is the mass of a cylinder of fluid with radius, b, and unit depth:

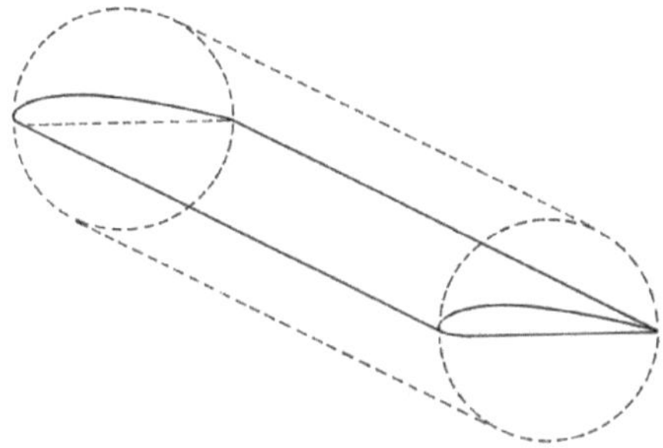

The second term on the right-hand side of Equation (6.5) is the circulatory term, in that it satisfies both the boundary conditions and the Kutta condition. Also, the shed wake is accounted for, as expressed by the $C(k)$ term, which is referred to as the "Theodorsen function," in honor of the NACA researcher who first derived these equations in 1935. $C(k)$ is a complex Bessel function, $C(k)=F(k)+iG(k)$:

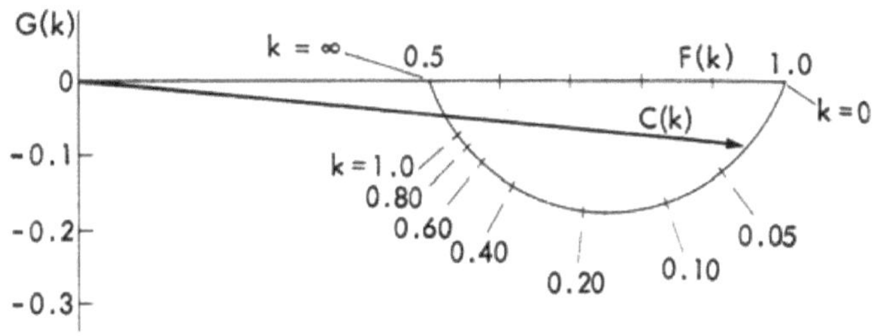

Approximate equations for $F(k)$ and $G(k)$, for $0.1 \leqslant k \leqslant 3.0$, are provided by A. I. Van de Vooren in the *AGARD Manual on Aeroelasticity*, vol. 6:

$$F(k)=\frac{0.021573+0.210400k+0.512607k^2+0.500502k^3}{0.021508+0.251239k+1.035378k^2+k^3}$$

$$G(k)=\frac{-0.001995-0.327214k-0.122397k^2-0.000146k^3}{0.089318+0.934530k+2.481481k^2+k^3}$$

A fundamental assumption for Theodorsen's analysis was that the shed wake is planar with the wing chord. However, for $k \gtrsim 0.4\ (\lambda \lesssim 8)$, theory and actuality diverge because the wake is no longer flat.

Now, consider the limiting case for $k \to 0\ (\omega \to 0,\ \lambda \to \infty)$. The previous figure shows that $C(k) \to 1$. Also, for this situation, $\dot{\alpha},\ \ddot{\alpha},\ \dot{h},\ \ddot{h} \to 0$, so Equation (6.5) becomes

$$L'_{k\to 0}=2\pi\rho V b(V\alpha)=2\pi\rho\frac{V^2}{2}(2b)\alpha$$

which is the two-dimensional airfoil steady lift equation.

Tables are available that give L', $M'_{c/4}$, and P' (which can be transformed to L', M', and P') for the two-dimensional airfoil, including control surfaces. For example, the *AGARD Manual on Aeroelasticity* (vol. 6) gives

$$L'=\pi\rho V^2 b\left(k_a\,\bar{h}_{c/4}/b+k_b\,\bar{\alpha}+k_c\,\bar{\beta}\right)e^{i\omega t}$$

$$M'_{c/4}=\pi\rho V^2 b^2\left(m_a\,\bar{h}_{c/4}/b+m_b\,\bar{\alpha}+m_c\,\bar{\beta}\right)e^{i\omega t}$$

$$P'=\pi\rho V^2 b^2\left(n_a\,\bar{h}_{c/4}/b+n_b\,\bar{\alpha}+n_c\,\bar{\beta}\right)e^{i\omega t}$$

where k_a, k_b, m_a, m_b, n_a, n_b are *complex* functions of k, and k_c, m_c, n_c are *complex* functions of k and the control size. Also, the Appendix of Scanlan and Rosenbaum's *Introduction to the Study of Aircraft Vibration and Flutter* (Macmillan Co., 1951), has tables for subsonic and supersonic speeds.

Finally, one may write the aerodynamic forcing terms as

$$\begin{Bmatrix} (F_h)_{aero} \\ (M_\alpha)_{aero} \\ (M_\beta)_{aero} \end{Bmatrix} = \begin{Bmatrix} -L' \\ -M' \\ -P' \end{Bmatrix} = [Aero]\begin{Bmatrix} h \\ \alpha \\ \beta \end{Bmatrix}$$

where *[Aero]* involves terms that are complex functions of k.

6.2.5 Complete Flutter Equations

The inertial, elastic, damping, and aerodynamic terms combine to give the complete set of equations for flutter of the representative section:

$$[Aero]\begin{Bmatrix} h \\ \alpha \\ \beta \end{Bmatrix} = [Inertial]\begin{Bmatrix} \ddot{h} \\ \ddot{\alpha} \\ \ddot{\beta} \end{Bmatrix} + [Damping]\begin{Bmatrix} \dot{h} \\ \dot{\alpha} \\ \dot{\beta} \end{Bmatrix} + [Elastic]\begin{Bmatrix} h \\ \alpha \\ \beta \end{Bmatrix}$$

This is a set of second-order linear differential equations with complex coefficients. Further, because the representations for damping and the aerodynamic terms are only valid for simple harmonic motions, the variables are constrained to have the form

$$h = \bar{h}e^{i\omega t},\ \alpha = \bar{\alpha}e^{i\omega t},\ \beta = \bar{\beta}e^{i\omega t}$$

The resulting characteristic equation then requires special methods for solution. By contrast, if damping is ignored and quasi-steady aerodynamics are assumed, one may solve this in the usual stability-root fashion by assuming

$$h = \bar{h}e^{\lambda t},\ \alpha = \bar{\alpha}e^{\lambda t},\ \beta = \bar{\beta}e^{\lambda t},\ \text{where } \lambda = \lambda_r + i\omega$$

In fact, this solution method may also have validity for the complete equations if the damping is light, as will be shown later.

The representative section may be related to the solution for a finite wing in the same fashion as was done for the divergence case.

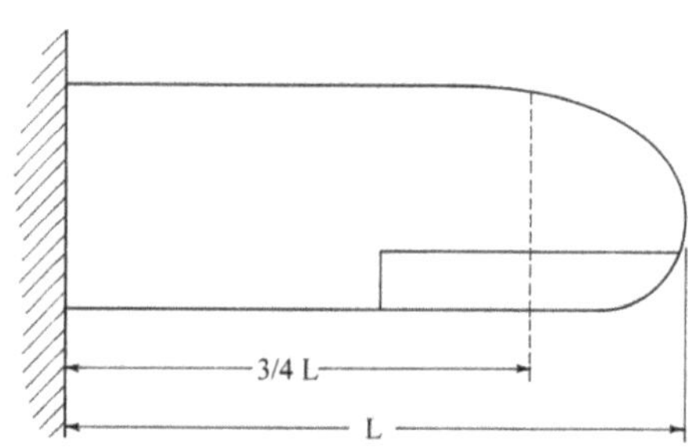

It will be shown later that choosing the parameters at the $3/4L$ spanwise location provides a reasonable approximation.

Finally, it has been found useful to express the elastic terms as functions of the *uncoupled* natural frequencies

$$\omega_h^2 = \frac{K_h}{M},\ \omega_\alpha^2 = \frac{K_\alpha}{I_\alpha},\ \omega_\beta^2 = \frac{K_\beta}{I_\beta}$$

so that one obtains

$$K_h = M\,\omega_h^2,\ K_\alpha = I_\alpha\,\omega_\alpha^2,\ K_\beta = I_\beta\,\omega_\beta^2$$

6.3 REPRESENTATIVE SECTION WITH NO CONTROL SURFACE

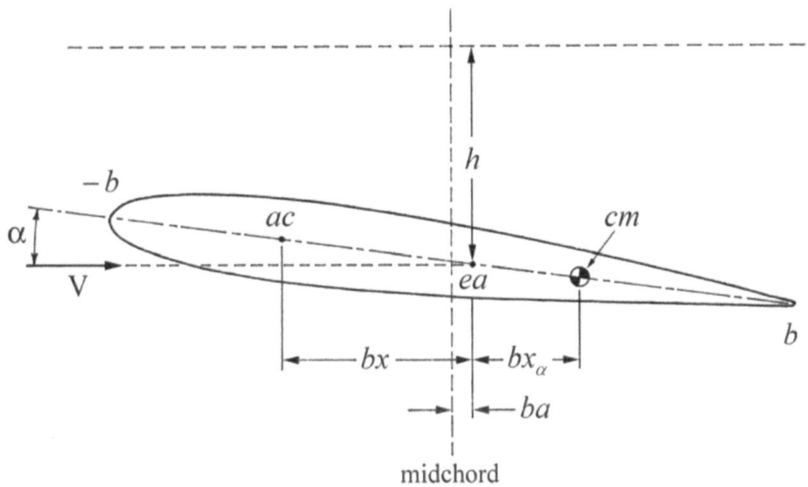

For this special case, the following assumptions will be made:

1. Structural damping is negligible, which gives $[Damping]=0$.
2. The aerodynamic terms may be represented by

$$L' = qSC_{L_\alpha}\alpha,\ M' = -bxqSC_{L_\alpha}\alpha$$

Therefore, the equations become

$$\begin{bmatrix} M & S_\alpha \\ S_\alpha & I_\alpha \end{bmatrix}\begin{Bmatrix} \ddot{h} \\ \ddot{\alpha} \end{Bmatrix} + \begin{bmatrix} M\omega_h^2 & 0 \\ 0 & I_\alpha\omega_\alpha^2 \end{bmatrix}\begin{Bmatrix} h \\ \alpha \end{Bmatrix} = \begin{bmatrix} 0 & -qSC_{L_\alpha} \\ 0 & bxqSC_{L_\alpha} \end{bmatrix}\begin{Bmatrix} h \\ \alpha \end{Bmatrix}$$

which expand to

$$M\ddot{h} + S_\alpha\ddot{\alpha} + M\omega_h^2 h = -qSC_{L_\alpha}\alpha$$

$$S_\alpha\ddot{h} + I_\alpha\ddot{\alpha} + I_\alpha\omega_\alpha^2\alpha = bxqSC_{L_\alpha}\alpha$$

Assume solutions of the form

$$h=\bar{h}\,e^{\lambda t},\ \alpha=\bar{\alpha}\,e^{\lambda t}$$

Upon substitution into the equations, one obtains

$$\begin{bmatrix} M\lambda^2+M\omega_h^2 & S_\alpha\lambda^2+qSC_{L_\alpha} \\ S_\alpha\lambda^2 & I_\alpha\lambda^2+I_\alpha\omega_\alpha^2-bxqSC_{L_\alpha} \end{bmatrix}\begin{Bmatrix}\bar{h}\\ \bar{\alpha}\end{Bmatrix}=0$$

With the determinant of the square matrix set equal to zero, one obtains the characteristic equation

$$\left(MI_\alpha-S_\alpha^2\right)\lambda^4+\left(M\omega_h^2I_\alpha+M\omega_\alpha^2I_\alpha-MbxqSC_{L_\alpha}-S_\alpha qSC_{L_\alpha}\right)\lambda^2$$
$$+M\omega_h^2\left(I_\alpha\omega_\alpha^2-bxqSC_{L_\alpha}\right)=0$$

Divide this equation by $M\,I_\alpha\omega_\alpha^4$. Also, note that

$$x_\alpha=\frac{S_\alpha}{Mb}\ \text{and}\ r_\alpha^2=\frac{I_\alpha}{Mb^2},\ \text{so that}\ \frac{S_\alpha^2}{MI_\alpha}=\left(\frac{x_\alpha}{r_\alpha}\right)^2$$

Further, define

$$Q\equiv\frac{qSC_{L_\alpha}b}{I_\alpha\omega_\alpha^2},\ \text{“Dynamic-Pressure Parameter”}$$

So the equation becomes

$$\left[1-\frac{x_\alpha^2}{r_\alpha^2}\right]\left(\frac{\lambda}{\omega_\alpha}\right)^4+\left[1+\left(\frac{\omega_h}{\omega_\alpha}\right)^2-Q\left(x+x_\alpha\right)\right]\left(\frac{\lambda}{\omega_\alpha}\right)^2+\left[1-Qx\right]\left(\frac{\omega_h}{\omega_\alpha}\right)^2=0 \qquad (6.6)$$

which is of the form

$$\hat{a}\left(\frac{\lambda}{\omega_\alpha}\right)^4+\hat{b}\left(\frac{\lambda}{\omega_\alpha}\right)^2+\hat{c}=0$$

which is a quadratic equation in $\left(\lambda/\omega_\alpha\right)^2$.

Note that $\hat{c}$ gives divergence information:

$$\hat{c}=0\rightarrow Qx=1\rightarrow q_{div}=\frac{K_\alpha}{SC_{L_\alpha}bx}\ \text{(recall that}\ bx=e)$$

Therefore, $\hat{c}=0$ gives a non-oscillatory real root which is equal to zero and is a divergence boundary. However, for no divergence $\hat{c}$ *is positive*, which gives that $q \prec q_{div}$.

Also, because

$$I_\alpha = Mb^2 r_\alpha^2 \succ Mb^2 x_\alpha^2, \text{ then } r_\alpha^2 \succ x_\alpha^2$$

and so $\hat{a}$ *is positive* and $\prec$ 1.

$\hat{b}$ is positive or negative depending on $x + x_\alpha$ and Q, so the solution, $(\lambda / \omega_\alpha)^2$, may be real or complex depending on the sign of $\hat{b}^2 - 4\hat{a}\hat{c}$ (the discriminant).

$$\hat{\mathbf{b}}^2 - 4\hat{\mathbf{a}}\hat{\mathbf{c}} \succ 0$$

Let $\delta \equiv \left(\hat{b}^2 - 4\hat{a}\hat{c}\right)^{1/2} \rightarrow$ real , then

$$\left(\frac{\lambda}{\omega_\alpha}\right)^2 = \frac{-\hat{b} \pm \delta}{2\hat{a}} \prec 0 \text{ and real}$$

and so one obtains

$$\left(\frac{\lambda}{\omega_\alpha}\right) = \pm i\left(\frac{\omega_1}{\omega_\alpha}\right), \pm i\left(\frac{\omega_2}{\omega_\alpha}\right)$$

This gives two distinct oscillatory roots. The real part of λ equals zero, so one has *neutral* stability.

$$\hat{\mathbf{b}}^2 - 4\hat{\mathbf{a}}\hat{\mathbf{c}} \prec 0$$

Let $\hat{u} = -\hat{b} / 2\hat{a}$, $v = \left(4\hat{a}\hat{c} - \hat{b}^2\right)^{1/2} / 2\hat{a}$, then

$$\left(\frac{\lambda}{\omega_\alpha}\right)^2 = \hat{u} \pm i\hat{v}$$

which gives

$$\left(\frac{\lambda}{\omega_\alpha}\right) = \pm\frac{\lambda_r}{\omega_\alpha} \pm i\frac{\omega}{\omega_\alpha}$$

There is a separation in damping, including an unstable value, but the frequencies are identical. The following figures show the behavior of the roots for the two cases. For example, Figure (a) shows how the ω_1 / ω_α and ω_2 / ω_α roots for case 1 coalesce to give the ω / ω_α frequency for case 2. Also, Figure (b) shows how the square root of $\hat{u} \pm i\hat{v}$ gives the four roots for case 2, $\pm\lambda_r / \omega_\alpha \pm i\omega / \omega_\alpha$:

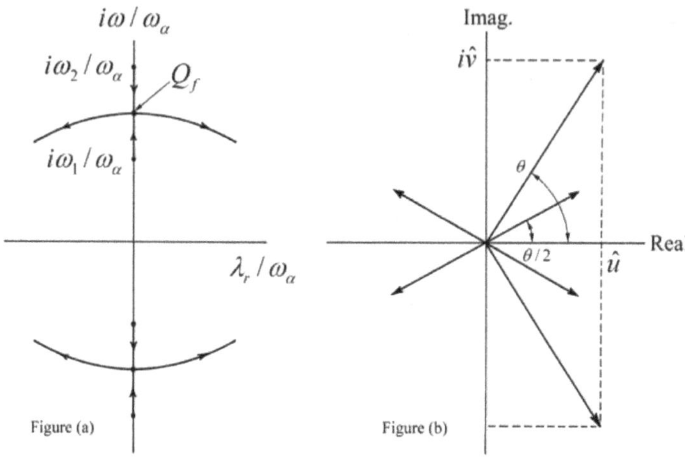

$$\hat{\mathbf{b}}^2 - 4\hat{\mathbf{a}}\hat{\mathbf{c}} = 0$$

The *flutter frequency* is defined to be given by $\hat{b}^2 - 4\hat{a}\hat{c} = 0 \rightarrow$

$$\left(\frac{\lambda}{\omega_\alpha}\right)^2 \equiv \left(\frac{\omega_f}{\omega_\alpha}\right)^2 = -\frac{\hat{b}}{2\hat{a}} = \left(\frac{\hat{c}}{\hat{a}}\right)^{\frac{1}{2}} = \left(\frac{\omega_h}{\omega_\alpha}\right)\left[\frac{1-Q_f x}{1-(x_\alpha/r_\alpha)^2}\right]^{1/2}$$

where Q_f gives the velocity, V_f, at which flutter occurs.

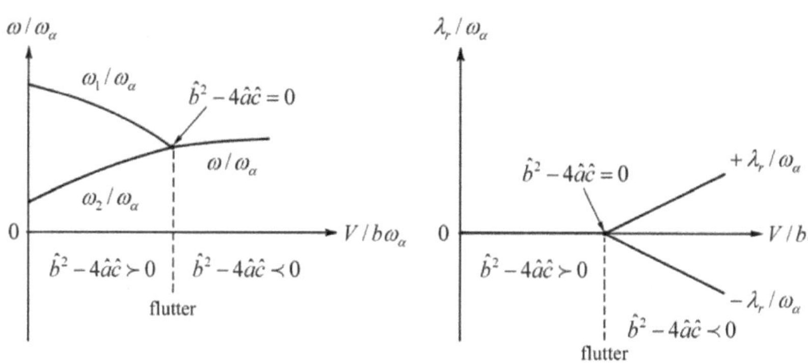

For this simplified model, the system is either neutrally stable or unstable. This is because there is no structural or material damping ($g = 0$) and no aerodynamic damping (given by $\dot{h}$ and $\dot{\alpha}$ terms). Therefore, the point at which instability first occurs has been defined to be the *flutter condition*. Note also, for this model, that this is the point where the ω_1 and ω_2 frequencies coalesce to a single value, ω.

The flutter condition, $\hat{b}^2 - 4\hat{a}\hat{c} = 0$, may be expanded by substituting the definitions for $\hat{a}$, $\hat{b}$, and $\hat{c}$:

$$(x+x_\alpha)^2 Q^2 - 2\left\{(x+x_\alpha)\left[1+\left(\frac{\omega_h}{\omega_\alpha}\right)^2\right] - 2\left(\frac{\omega_h}{\omega_\alpha}\right)^2 x\left[1-\left(\frac{x_\alpha}{r_\alpha}\right)^2\right]\right\}Q + \left[1+\left(\frac{\omega_h}{\omega_\alpha}\right)^2\right]^2$$
$$-4\left[1-\left(\frac{x_\alpha}{r_\alpha}\right)^2\right]\left(\frac{\omega_h}{\omega_\alpha}\right)^2 = 0$$

This is a quadratic equation for Q, from which the solutions are

$$Q_{1,2} = \frac{1}{(x+x_\alpha)^2}\left\{(x+x_\alpha)\left[1+\left(\frac{\omega_h}{\omega_\alpha}\right)^2\right] - 2x\left(\frac{\omega_h}{\omega_\alpha}\right)^2\left[1-\left(\frac{x_\alpha}{r_\alpha}\right)^2\right]\right\}$$
$$\pm\frac{2}{(x+x_\alpha)^2}\left[x_\alpha\left(\frac{\omega_h}{\omega_\alpha}\right)^2\left[1-\left(\frac{x_\alpha}{r_\alpha}\right)^2\right]\left\{(x+x_\alpha) - x\left(\frac{\omega_h}{\omega_\alpha}\right)^2\left(1+\frac{x x_\alpha}{r_\alpha^2}\right)\right\}\right]^{1/2} \tag{6.7}$$

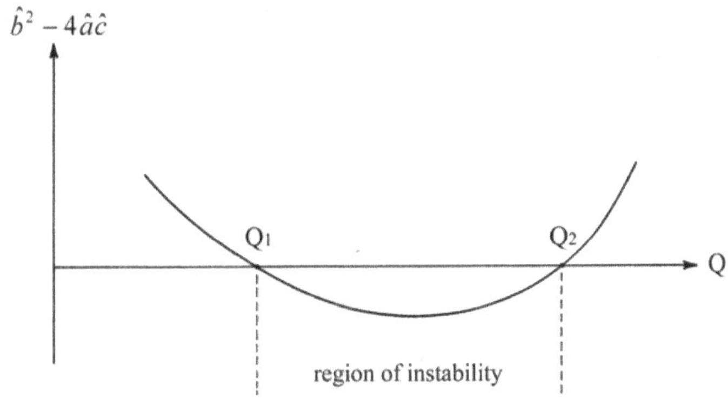

From Equation (6.7), for Q to be real, the $[\ldots]^{1/2}$ term must be real; therefore,

$$\left[x_\alpha\left(\frac{\omega_h}{\omega_\alpha}\right)^2\left[1-\left(\frac{x_\alpha}{r_\alpha}\right)^2\right]\left\{(x+x_\alpha) - x\left(\frac{\omega_h}{\omega_\alpha}\right)^2\left(1+\frac{x x_\alpha}{r_\alpha^2}\right)\right\}\right]^{1/2} \geqslant 0$$

Note that

$$\left(\frac{\omega_h}{\omega_\alpha}\right)^2 \geqslant 0, \text{and } \left[1-\left(\frac{x_\alpha}{r_\alpha}\right)^2\right] \geqslant 0$$

This leaves

$$x_\alpha\left\{(x+x_\alpha) - x\left(\frac{\omega_h}{\omega_\alpha}\right)^2\left(1+\frac{x x_\alpha}{r_\alpha^2}\right)\right\} \geqslant 0 \tag{6.8}$$

which is the condition for Q to be real.

For Q to be positive, the first {. . .} term in Equation (6.7) must be $\geqslant 0$, which gives that

$$x + x_\alpha + \left(\frac{\omega_h}{\omega_\alpha}\right)^2 \left[x_\alpha - x + 2x\left(\frac{x_\alpha}{r_\alpha}\right)^2\right] \geqslant 0 \tag{6.9}$$

6.3.1 Aside

Note that if Equation (6.7) is written as $Q_{1,2} = A \pm B$, one may readily show that $|A| \geqslant |B|$.

6.3.2 End Aside

Conclusions may be reached about the conditions for flutter from Equations (6.8) and (6.9), for this analytical model. This exercise was developed by S. Pines ("An Elementary Explanation of the Flutter Mechanism," *Proc.* Nat. Specialists Meeting on Dynamics and Aeroelasticity, Institute of the Aeronautical Sciences, Ft. Worth, Texas, November 1958, pp. 52–58).

Case 1: If the *cm* is ahead of the $ea(x_\alpha \prec 0)$, then flutter is impossible.

Proof:

From Condition (6.8), if $x_\alpha \prec 0$, then the {. . .} term must be < 0 for flutter to exist. If so, this gives

$$-x - x_\alpha + x\left(\frac{\omega_h}{\omega_\alpha}\right)^2 \left(1 + \frac{xx_\alpha}{r_\alpha^2}\right) \geqslant 0 \tag{6.10}$$

Now, upon combining Conditions (6.9) and (6.10), one obtains

$$x\left(\frac{\omega_h}{\omega_\alpha}\right)^2 \left(1 + \frac{xx_\alpha}{r_\alpha^2}\right) + \left(\frac{\omega_h}{\omega_\alpha}\right)^2 \left[x_\alpha - x + 2x\left(\frac{x_\alpha}{r_\alpha}\right)^2\right] = \left(\frac{\omega_h}{\omega_\alpha}\right)^2 \left[\frac{x^2 x_\alpha}{r_\alpha^2} + x_\alpha + 2x\left(\frac{x_\alpha}{r_\alpha}\right)^2\right] \geqslant 0 \tag{6.11}$$

However, because

$$\left(\frac{\omega_h}{\omega_\alpha}\right)^2 \geqslant 0$$

then Condition (6.11) becomes

$$\left(x^2 + r_\alpha^2\right)x_\alpha + 2xx_\alpha^2 \geqslant 0 \rightarrow x_\alpha\left[2xx_\alpha + \left(x^2 + r_\alpha^2\right)\right] \geqslant 0 \tag{6.12}$$

Further, one has that

$$\left(x^2 + r_\alpha^2\right) \geqslant 0, \qquad \text{and } x_\alpha^2 \geqslant 0$$

so if $x_\alpha \prec 0$ and $x \prec 0$, then Condition (6.12) is actually $\leqslant 0$ and flutter cannot exist (Condition (6.11) is violated).

For the situation where $x \succ 0$ and $x_\alpha \prec 0$, one sees from the right-hand side of Condition (6.12) that

$$2xx_\alpha + \left(x^2 + r_\alpha^2\right) = -2x|x_\alpha| + \left(x^2 + r_\alpha^2\right) \succ -2xr_\alpha + \left(x^2 + r_\alpha^2\right) = \left(x - r_\alpha\right)^2 \succ 0$$

Therefore, once again, Condition (6.12) is $\leqslant 0$ and flutter cannot exist.

Case 2: If the *ac* lies between the *ea* and the *cm*, then flutter is always possible.

Proof:

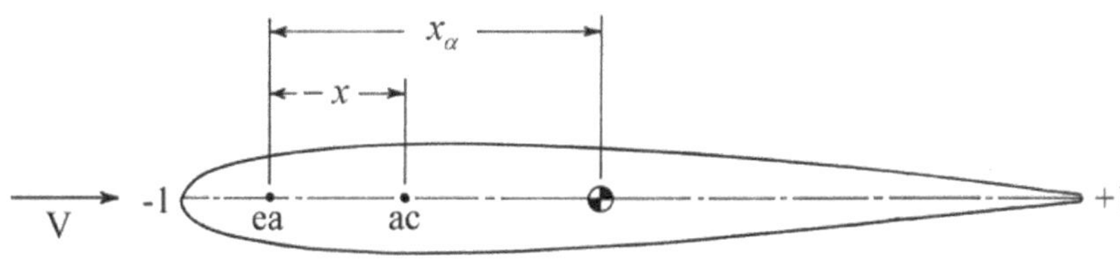

$$x_\alpha \succ 0,\ x \prec 0,\ x_\alpha \succ |x|$$

From Condition (6.8), one has that

$$x_\alpha \left\{ \left(x + x_\alpha\right) - x\left(\frac{\omega_h}{\omega_\alpha}\right)^2 \left(1 + \frac{xx_\alpha}{r_\alpha^2}\right) \right\} = (+)\{(+) - (-)(+)(+)\} \geqslant 0$$

So that condition is satisfied. Next, from Condition (6.9), one has that

$$\left(x + x_\alpha\right) + \left(\frac{\omega_h}{\omega_\alpha}\right)^2 \left[\left(x_\alpha - x\right) + 2x\left(\frac{x_\alpha}{r_\alpha}\right)^2\right] = (+) + (+)[(+) + (-)] = (+) + (+)[(+)] \geqslant 0$$

Therefore, the second condition is satisfied, which means that there is some velocity at which flutter will occur for this case.

Case 3: If the *ac* is forward of the *ea*, then flutter is possible only if

$$\left(\frac{\omega_h}{\omega_\alpha}\right)^2 \leqslant \frac{x + x_\alpha}{x\left(1 + \frac{xx_\alpha}{r_\alpha^2}\right)}$$

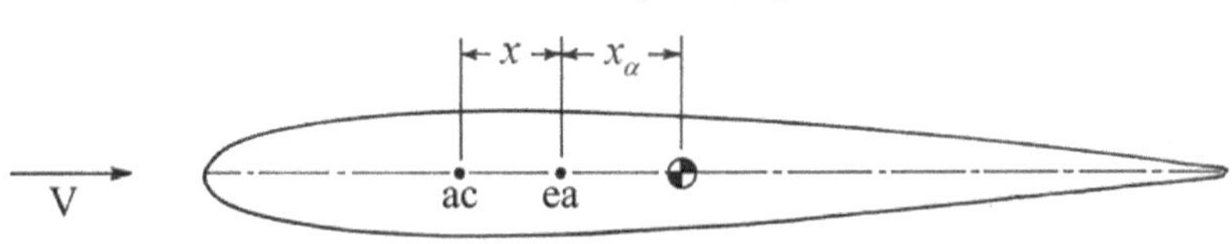

$$x_\alpha \succ 0,\ x \succ 0$$

Proof:

Because $x_\alpha \succ 0$, then Condition (6.8) becomes {. . .} ⩾ 0, which gives the preceding result. Also, this may be substituted into Condition (6.9), which gives that

$$(x+x_\alpha)\left[x\frac{xx_\alpha}{r_\alpha^2}+x_\alpha+2x\left(\frac{x_\alpha}{r_\alpha}\right)^2\right]\geqslant 0$$

Therefore, Condition (6.9) is satisfied.

Case 4: If the *ac* is aft of the *cm*, then flutter is possible only if

$$\left(\frac{\omega_h}{\omega_\alpha}\right)^2 \geqslant \frac{x+x_\alpha}{x\left(1+\frac{xx_\alpha}{r_\alpha^2}\right)}, \quad \text{and} \quad r_\alpha^2 \succ x_\alpha |x|$$

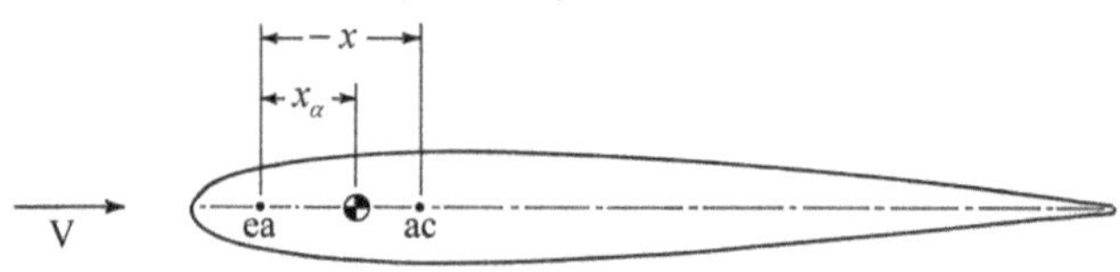

$x_\alpha \succ 0,\ x \prec 0,\ |x| \succ x_\alpha$

Proof:

$x_\alpha \succ 0$, so from Condition (6.8), one has that {. . .} ⩾ 0 for flutter to exist. This gives that

$$-x-x_\alpha+x\left(\frac{\omega_h}{\omega_\alpha}\right)^2\left(1+\frac{xx_\alpha}{r_\alpha^2}\right)\leqslant 0$$

However, x is a negative value for this situation, so $-x=|x|$, from which one obtains that

$$|x|-x_\alpha \leqslant |x|\left(\frac{\omega_h}{\omega_\alpha}\right)^2\left(1+\frac{xx_\alpha}{r_\alpha^2}\right)$$

from which the condition is obtained.

6.3.3 Graphical Summary

The following figures show the flutter condition for various *cm* locations. The first figure is for the case where the *ac* is aft of the *ea*:

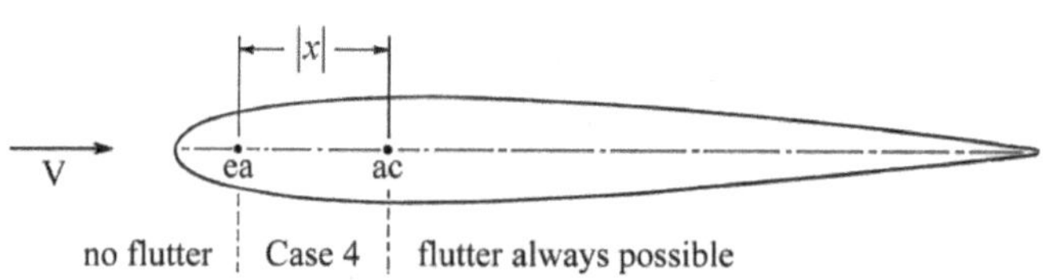

The second figure shows what happens when the *ea* is aft of the *ac* (which is usually the case for airplane wings):

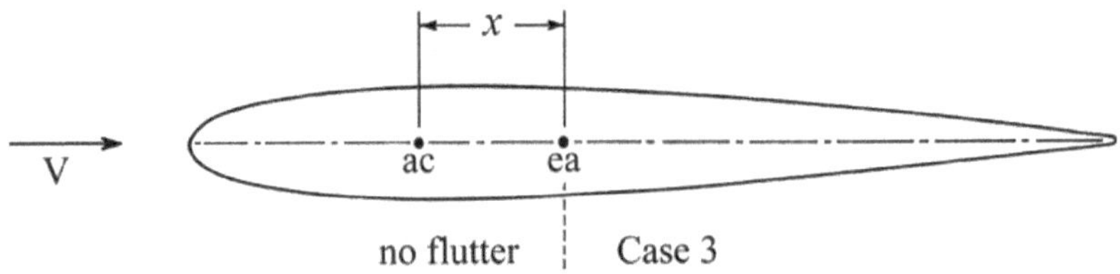

For both cases, as originally stated in case 1, when the mass center is forward of the elastic axis, no flutter is possible. However, it must be emphasized that these results are for a very simple analytical model, with no damping and quasi-steady aerodynamic forcing that is a function of α only.

Other interesting results may be obtained from Pine's analytical model. For example, if $(\omega_h / \omega_\alpha)$ is small, then Equation (6.7) reduces to

$$Q_{1,2} \simeq \frac{1}{x + x_\alpha} \pm \frac{2(\omega_h / \omega_\alpha)\left\{x_\alpha\left[1-(x_\alpha / r_\alpha)^2\right]\right\}^{1/2}}{(x_\alpha + x)^{3/2}} \tag{6.13}$$

For flutter to exist, the following two conditions must be met:

$$(x + x_\alpha) \geqslant 0, \quad x_\alpha \geqslant 0$$

This ensures that the $1/(x + x_\alpha)$ and the {. . .} terms are positive. Also, this result states that the *cm* must be aft of both the *ea* and *ac* for flutter.

Now, reduce the frequency ratio even further, $(\omega_h / \omega_\alpha) \ll 1$, then Equation (6.13) becomes

$$Q_{flutter} = \frac{q_f SbC_{L_\alpha}}{I_\alpha \omega_\alpha^2} \approx \frac{1}{x + x_\alpha}$$

From $I_\alpha = mb^2 r_\alpha^2$, this gives that

$$\frac{V_f}{b\omega_\alpha} \approx \left[\left(\frac{2m}{\rho bS}\right)\frac{r_\alpha^2}{C_{L_\alpha}(x + x_\alpha)}\right]^{1/2}$$

In the limit, when $\omega_h \approx 0$, Equation (6.13) becomes

$$Q = \frac{1}{x + x_\alpha}$$

which is a state of *dynamic divergence*:

$$q_{dd} = \frac{I_\alpha \omega_\alpha^2}{SC_{L_\alpha} b(x_\alpha + x)} = \frac{K_\alpha}{SC_{L_\alpha} b(x_\alpha + x)}$$

Compare this with the *static* divergence equation:

$$q_{sd} = \frac{1}{SC_{L_\alpha} F_\theta e} = \frac{K_\theta}{SC_{L_\alpha} e}$$

$K_\theta = K_\alpha$, but e is replaced by $b(x_\alpha + x)$. This implies that an inertial reaction moment, caused by the *cm* not being coincident with the *ea*, will have an effect during a pitching divergence acceleration.

6.3.4 Numerical Examples

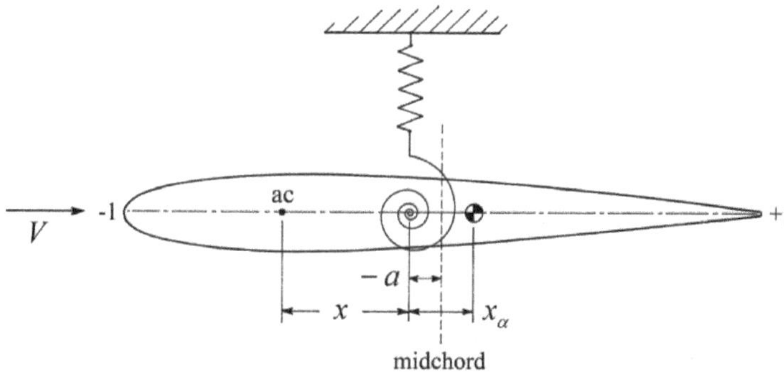

The non-flapped section has the following properties:

$$a = -0.1,\ x_\alpha = 0.2,\ x = a + 0.5 = 0.4,\ b = 1.0,\ \omega_\alpha = 1.0,\ \left(\omega_h / \omega_\alpha\right)^2 = 0.25$$

The lift-curve slope of the section is $C_{L_\alpha} = 2\pi$, and the reference area, S, is chosen to be $c \times (1) = 2b = 2.0$. Further, the mass, M, is given by

$$M = 10.0\pi\rho b^2$$

and the square of the non-dimensional radius of gyration is $r_\alpha^2 = 0.25$. With this information, one obtains from Equation (6.7) that

$$Q_{1,2} = 1.61668 \pm 0.77889 \rightarrow Q_{flutter} = 0.83779$$

Further, from the definition of Q, one has that

$$Q_{flutter} = \frac{q_f S C_{L_\alpha} b}{I_\alpha \omega_\alpha^2} = \frac{\rho V_f^2}{2} \frac{(2b)(2\pi)(b)}{M r_\alpha^2 b^2 \omega_\alpha^2} = \frac{2\pi\rho V_f^2}{M r_\alpha^2 \omega_\alpha^2} = \frac{8\pi\rho V_f^2}{M\omega_\alpha^2}$$

Upon substituting the defined value for M, one obtains the flutter speed:

$$Q_{flutter} = \frac{0.8 V_f^2}{b^2 \omega_\alpha^2} = 0.83779 \rightarrow \frac{V_f}{b\omega_\alpha} = 1.0233$$

If $V/b\omega_\alpha$ is allowed to vary, one may find the real and imaginary roots of Equation (6.6) and plot these as shown:

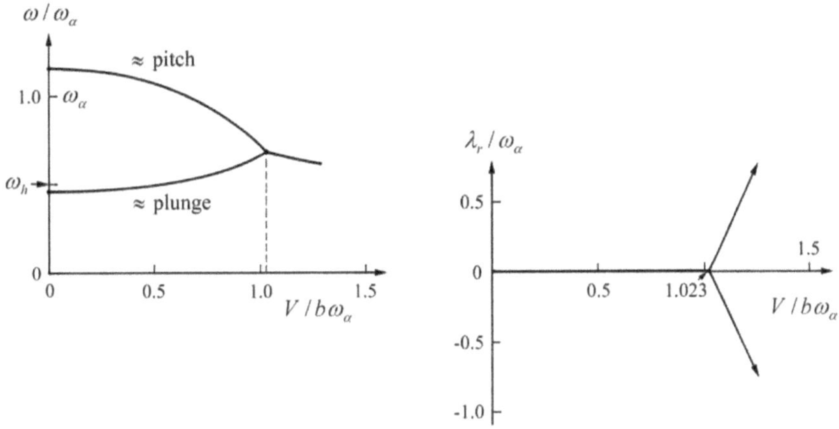

Next, consider a solution that accounts for plunging, $\dot{h}$, in the aerodynamic equations:

$$L' = 2\pi\rho V^2 b\left(\alpha + \dot{h}/V\right),\ M' = -2\pi\rho V^2 b^2 x\left(\alpha + \dot{h}/V\right)$$

The dynamic equations become

$$\begin{bmatrix} M & S_\alpha \\ S_\alpha & I_\alpha \end{bmatrix}\begin{Bmatrix} \ddot{h} \\ \ddot{\alpha} \end{Bmatrix} + \begin{bmatrix} M\omega_h^2 & 0 \\ 0 & I_\alpha\omega_\alpha^2 \end{bmatrix}\begin{Bmatrix} h \\ \alpha \end{Bmatrix} = 2\pi\rho V^2 b\begin{bmatrix} 0 & -1 \\ 0 & bx \end{bmatrix}\begin{Bmatrix} h \\ \alpha \end{Bmatrix} + 2\pi\rho Vb\begin{bmatrix} -1 & 0 \\ bx & 0 \end{bmatrix}\begin{Bmatrix} \dot{h} \\ \dot{\alpha} \end{Bmatrix}$$

Upon assuming a solution of the form

$$h = \bar{h}e^{\lambda t}, \alpha = \bar{\alpha}e^{\lambda t}$$

one may obtain the characteristic equation for λ:

$$\left(MI_\alpha - S_\alpha^2\right)\lambda^4 + 2\pi\rho Vb\left(I_\alpha + bxS_\alpha\right)\lambda^3 + \left[(MI_\alpha\left(\omega_h^2 + \omega_\alpha^2\right) - 2\pi\rho V^2 b\left(Mbx + S_\alpha\right)\right]\lambda^2$$
$$+\ 2\pi\rho VbI_\alpha\omega_\alpha^2\lambda + M\omega_h^2\left(I_\alpha\omega_\alpha^2 - 2\pi\rho V^2 b^2 x\right) = 0$$

As done with the previous model, this may be non-dimensionalized to give

$$\tilde{a}\left(\frac{\lambda}{\omega_\alpha}\right)^4 + \tilde{b}\left(\frac{\lambda}{\omega_\alpha}\right)^3 + \tilde{c}\left(\frac{\lambda}{\omega_\alpha}\right)^2 + \tilde{d}\left(\frac{\lambda}{\omega_\alpha}\right) + \tilde{e} = 0$$

The roots of this are

$$\frac{\lambda}{\omega_\alpha} = \frac{\lambda_r}{\omega_\alpha} + i\frac{\omega}{\omega_\alpha}$$

and for the numerical example, these plot out vs. $V / b\omega_\alpha$ as shown:

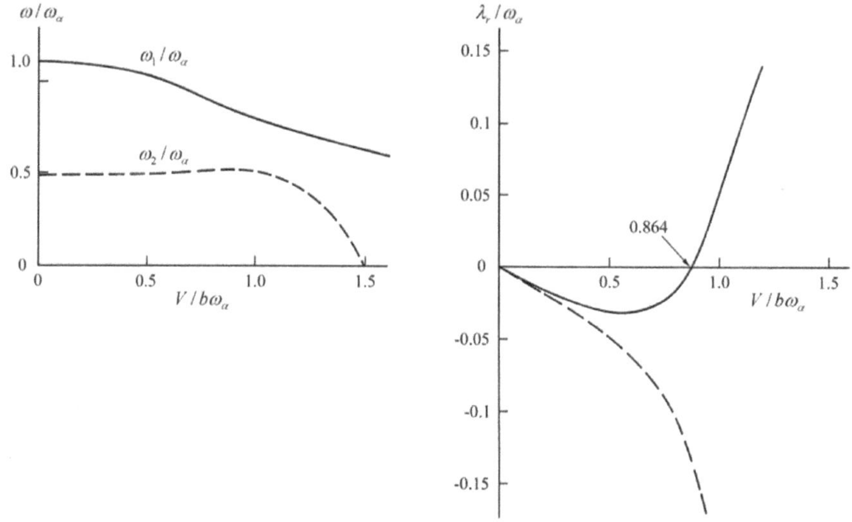

Unlike the simpler model, the motion is either stable or unstable, with flutter occurring at

$$\frac{V_f}{b\omega_\alpha} = 0.86$$

This is due to the aerodynamic damping provided by the $\dot{h}$ term. Further, the two frequencies of oscillation remain distinct (do not coalesce). As with the simpler model, the higher-frequency oscillation is associated with a modal vector that is primarily pitching motion. In fact, this is the mode that goes unstable at the flutter condition. In a way, this serves to justify the simpler aerodynamic model that is a function only of pitch angle, α.

Next, consider the full aerodynamic expressions, as discussed earlier in this chapter:

$$L' = L'_{app} + L'_{circ} = \pi\rho b^2\left(\ddot{h} + V\dot{\alpha} - ba\ddot{\alpha}\right) + 2\pi\rho VbC(k)\left[V\alpha + \dot{h} + b\left(\frac{1}{2} - a\right)\dot{\alpha}\right] \quad (6.14)$$

$$\begin{aligned} M' = M'_{app} + M'_{circ} &= -\pi\rho b^2\left[ba\ddot{h} - Vb(1/2 - a)\dot{\alpha} - b^2\left(1/8 + a^2\right)\ddot{\alpha}\right] \\ &\quad - 2\pi\rho Vb^2 xC(k)\left[V\alpha + \dot{h} + b(1/2 - a)\dot{\alpha}\right] \end{aligned} \quad (6.15)$$

Where the $()_{app}$ terms account for apparent-mass effects and the $()_{circ}$ terms account for the unsteady wake effects through the complex "Theodorsen function," $C(k)$, which is only *exactly* valid for simple-harmonic motion:

$$\alpha = \bar{\alpha}e^{i\omega t}, \; h = \bar{h}e^{i\omega t} \quad (6.16)$$

This fact will be used to cast the $()_{circ}$ terms into real form. First of all, upon including $C(k)=F(k)+iG(k)$ into L'_{circ}, one obtains

$$\begin{aligned} L'_{circ} &= 2\pi\rho VbF(k)\left[V\alpha+\dot{h}+b(1/2-a)\dot{\alpha}\right] \\ &\quad +2\pi\rho VbG(k)\left[Vi\alpha+i\dot{h}+b(1/2-a)i\dot{\alpha}\right] \end{aligned} \tag{6.17}$$

Next, Equations (6.16) give that

$$\dot{\alpha}=i\omega\bar{\alpha}e^{i\omega t}\text{, so that } i\alpha=i\bar{\alpha}e^{i\omega t}\rightarrow i\alpha=\frac{\dot{\alpha}}{\omega}=\frac{b\dot{\alpha}}{(Vk)}$$

and

$$\dot{h}=i\omega\bar{h}e^{i\omega t}\text{, so that } i\dot{h}=i\left(i\omega\bar{h}e^{i\omega t}\right)=-\omega\bar{h}e^{i\omega t}=-\omega h=-(Vk/b)h$$

Likewise,

$$i\dot{\alpha}=-(Vk/b)\alpha$$

Thus, Equation (6.17) becomes

$$\begin{aligned} L'_{circ} &= 2\pi\rho V^2b\left[F(k)-G(k)(1/2-a)k\right]\alpha-2\pi\rho V^2G(k)kh+2\pi\rho VbF(k)\dot{h} \\ &\quad + 2\pi\rho Vb^2\left[F(k)(1/2-a)+G(k)/k\right]\dot{\alpha} \end{aligned}$$

Next, consider L'_{app}. The double derivatives may be expressed as

$$\ddot{h}=-\omega^2\bar{h}e^{i\omega t}=-\omega^2h=-(Vk/b)^2h,\ \ddot{\alpha}=-(Vk/b)^2\alpha$$

so that one obtains

$$L'_{app}=-\rho V^2\left(\pi k^2\right)h+\rho V\left(\pi b^2\right)\dot{\alpha}+\rho V^2\left(\pi bak^2\right)\alpha$$

Therefore, upon defining

$$\begin{aligned} A'_{hh} &\equiv -2\pi G(k)k-\pi k^2,\ A'_{h\alpha}\equiv 2\pi b\left[F(k)-G(k)(1/2-a)k\right]+\pi bak^2 \\ A''_{hh} &\equiv 2\pi bF(k),\ A''_{h\alpha}\equiv 2\pi b^2\left[F(k)(1/2-a)+G(k)/k\right]+\pi b^2 \end{aligned}$$

Equation (6.14) becomes

$$L'=\rho V^2A'_{hh}h+\rho V^2A'_{h\alpha}\alpha+\rho VA''_{hh}\dot{h}+\rho VA''_{h\alpha}\dot{\alpha} \tag{6.18}$$

One may perform the same exercise on the equation for M' to obtain

$$M'=\rho V^2A'_{\alpha h}h+\rho V^2A'_{\alpha\alpha}\alpha+\rho VA''_{\alpha h}\dot{h}+\rho VA''_{\alpha\alpha}\dot{\alpha} \tag{6.19}$$

where

$$A'_{\alpha h} \equiv xb\left[2\pi G(k)k\right] + \pi bak^2,\; A'_{\alpha\alpha} \equiv -xb\left\{2\pi b\left[F(k) - G(k)(1/2 - a)k\right]\right\} - \pi b^2\left(1/8 + a^2\right)k^2$$

$$A''_{\alpha h} \equiv -xb\left[2\pi bF(k)\right],\; A''_{\alpha\alpha} \equiv -xb\left\{2\pi b^2\left[F(k)(1/2 - a) + G(k)/k\right]\right\} + \pi b^3(1/2 - a)$$

Equations (6.18) and (6.19) may be expressed in a matrix form:

$$\begin{Bmatrix} L' \\ M' \end{Bmatrix} = \rho V^2 \begin{bmatrix} A'_{hh} & A'_{h\alpha} \\ A'_{\alpha h} & A'_{\alpha\alpha} \end{bmatrix} \begin{Bmatrix} h \\ \alpha \end{Bmatrix} + \rho V \begin{bmatrix} A''_{hh} & A''_{h\alpha} \\ A''_{\alpha h} & A''_{\alpha\alpha} \end{bmatrix} \begin{Bmatrix} \dot{h} \\ \dot{\alpha} \end{Bmatrix} \equiv \tag{6.20}$$

$$\rho V^2 \left[A_S\right] \begin{Bmatrix} h \\ \alpha \end{Bmatrix} + \rho V \left[A_D\right] \begin{Bmatrix} \dot{h} \\ \dot{\alpha} \end{Bmatrix} \tag{6.21}$$

where $\left[A_S\right]$ may be interpreted as an aerodynamic "spring" matrix and $\left[A_D\right]$ may be interpreted as an aerodynamic "damping" matrix.

Now, Equations (6.21) are incorporated into the section's dynamic equations to give

$$\begin{bmatrix} M & S_\alpha \\ S_\alpha & I_\alpha \end{bmatrix} \begin{Bmatrix} \ddot{h} \\ \ddot{\alpha} \end{Bmatrix} + \begin{bmatrix} M\omega_h^2 & 0 \\ 0 & I_\alpha \omega_\alpha^2 \end{bmatrix} \begin{Bmatrix} h \\ \alpha \end{Bmatrix} = -\rho V^2 \left[A_S\right] \begin{Bmatrix} h \\ \alpha \end{Bmatrix} - \rho V \left[A_D\right] \begin{Bmatrix} \dot{h} \\ \dot{\alpha} \end{Bmatrix} \tag{6.22}$$

and as before, one may assume a solution of the form

$$h = \bar{h}\, e^{\lambda t},\; \alpha = \bar{\alpha}\, e^{\lambda t}$$

from which a characteristic equation for the stability roots, $\lambda = \lambda_r + i\omega$, is obtained. Note that the coefficients in this equation will be functions of k. Therefore, an iterative solution is required where a trial value of k is chosen, and from the resulting roots, a value of ω_f is obtained. From this, a new value of k is calculated. This new value is cycled through the characteristic equation to find yet a newer solution for k, and the process is repeated until convergence.

Solutions for the numerical example are shown as follows:

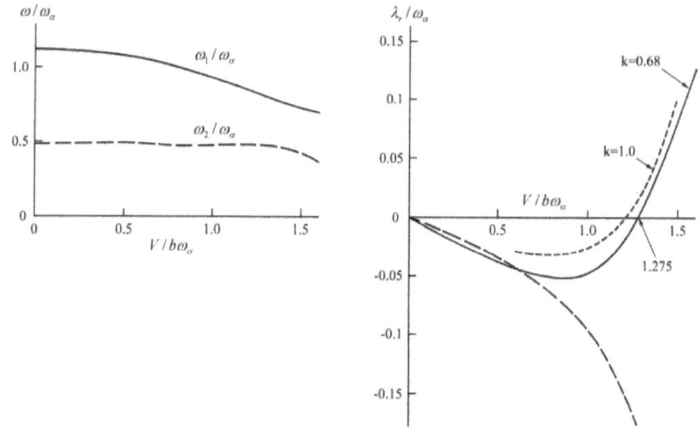

The first trial value was $k = 1$. The roots were then plotted vs. $V / b\omega_\alpha$, and a flutter value of $k = 0.68$ was found. Next, this value of k was tried, which gave a solution of $k = 0.64$. This was considered to be sufficiently converged to be the flutter solution, with the corresponding flutter speed parameter being

$$\frac{V_f}{b\omega_\alpha} = 1.27$$

A summary of all three solutions is given by the following table:

	Simple Model	**α and $\dot{h}$ Terms**	**Theodorsen**	**Theodorsen**
Assumed k	0.0	0.0	1.0	0.68
$V_f / b\omega_\alpha$	1.02	0.86	1.21	1.27
ω_f / ω_α	0.65	0.86	0.83	0.81
$k_{flutter}$	0.62	1.00	0.68	0.64
Flutter mode	$\approx$ *pitch*	$\approx$ *pitch*	$\approx$ *pitch*	$\approx$ *pitch*

It is seen that the shape of the $\lambda_r / b\omega_\alpha$ vs. $V / b\omega_\alpha$ curves for the last two examples bears a resemblance to the $V - g$ curves shown at the beginning of this chapter. However, although these curves give flutter information, they have fundamentally different interpretations. That is, the numerical examples just described have no structural damping whatsoever. Any damping at all comes from the aerodynamic terms (such as from $\dot{h}$ and $\dot{\alpha}$). The solutions took on the classic form of a stability problem with a characteristic equation and complex roots, the real parts giving modal damping information and the imaginary parts giving frequencies of oscillation. The crossover point where the real part of the root goes from stable to unstable is defined to be the flutter condition. In fact, this is the only point where the aerodynamic equations are exactly correct. For the other values of $V / b\omega_\alpha$, the curves may be assumed to be accurate only for lightly-damped oscillatory motions ($\lambda / \omega_\alpha \ll 1$) that are close to simple harmonic.

By comparison, every point on a $V - g$ curve is a flutter condition, with that value of g being the amount of structural damping required (positive or negative) to cause flutter. This is predetermined by the way in which the dynamic equations are solved, where the assumed solutions are undamped simple-harmonic motions (Equations 6.16). For a given V, there is an iteration involved to reach convergence on k. When that is achieved, the corresponding value of g that satisfies the equations gives a point on the $V - g$ curve. This is repeated for a full range of V values to obtain the complete curve.

6.3.5 Further Examples from Aeroelasticity Literature

Other solution examples for a non-flapped section may be obtained from *Principles of Aeroelasticity* by Bisplinghoff and Ashley, John Wiley and Sons, 1962, pp. 246–249. Incompressible aerodynamics are assumed, although the full Theodorsen equations are used.

6.3.5.1 Case 1: The Effect of Static Unbalance, x_α

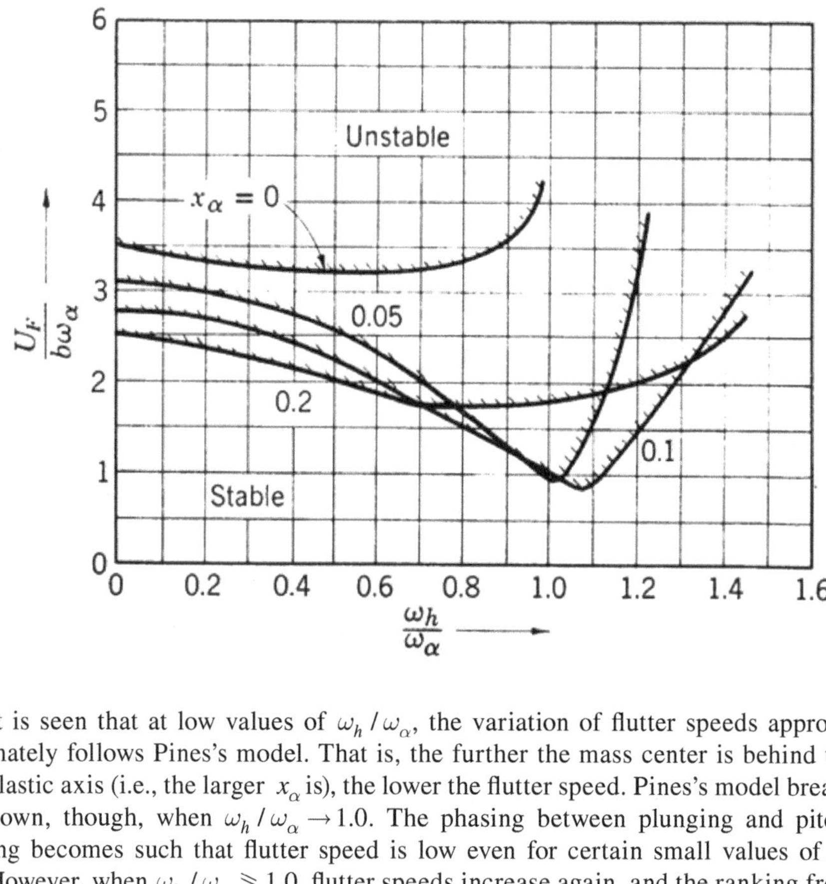

It is seen that at low values of ω_h / ω_α, the variation of flutter speeds approximately follows Pines's model. That is, the further the mass center is behind the elastic axis (i.e., the larger x_α is), the lower the flutter speed. Pines's model breaks down, though, when $\omega_h / \omega_\alpha \to 1.0$. The phasing between plunging and pitching becomes such that flutter speed is low even for certain small values of x_α. However, when $\omega_h / \omega_\alpha \gg 1.0$, flutter speeds increase again, and the ranking from Pines's model is restored.

6.3.5.2 Case 2: The Effect of Distance between the Mass Center and the Aerodynamic Center, $1/2 + a + x_\alpha = x + x_\alpha$

First of all, consider the case where $\omega_h / \omega_\alpha = 0$:

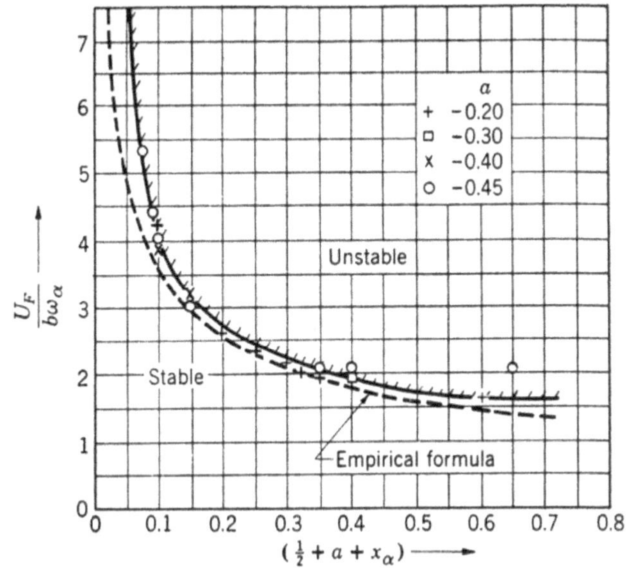

This is the dynamic divergence situation discussed earlier. In fact, the complete solution closely follows the equation from the simple model:

$$q_{dd} = \frac{I_\alpha^2 \omega_\alpha^2}{SC_{L_\alpha} b(x_\alpha + x)} \rightarrow \frac{U_{dd}}{b\omega_\alpha} = \left[\left(\frac{2m}{\pi \rho b S} \right) \frac{r_\alpha^2}{1 + 2(a + x_\alpha)} \right]^{1/2}$$

Note how the result depends on $1/2 + a + x_\alpha$ but is independent of a. Therefore, it is strictly a function of the distance between *ac* and *cm*.

Next, consider the results when $\omega_h / \omega_\alpha = 0.707$:

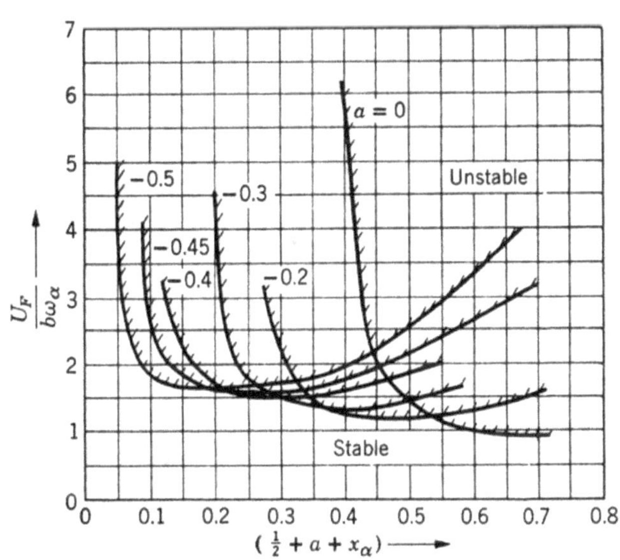

Note that for a given a, the cm moves aft with increasing $(1/2+a+x_\alpha)=(x+x_\alpha)$. Also, as a is decreased, x_α is increased:

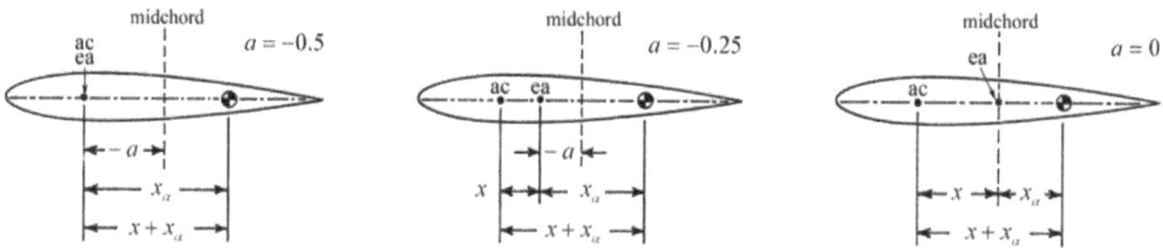

Three examples are shown in the preceding figure, where the distance between *ac* and*cm* is fixed. However, the flutter boundary is very dependent on the value of a. For example, looking from left to right on the $U_f/b\omega_\alpha$ vs. $(1/2+a+x_\alpha)$ plots, one sees that when $a=-0.5$, the flutter boundary is encountered at $(1/2+a+x_\alpha)\approx 0.05$. When a increases to −0.25, the flutter boundary increases to $(1/2+a+x_\alpha)\approx 0.25$. Finally, when $a=0$, the flutter boundary is at $(1/2+a+x_\alpha)\approx 0.4$. This trend serves to confirm Pines's result that the section's susceptibility to flutter decreases as x_α is decreased.

6.3.5.3 Case 3: The Effect of Mass Ratio, $\mu = \dfrac{2m}{(\pi\rho bS)}$

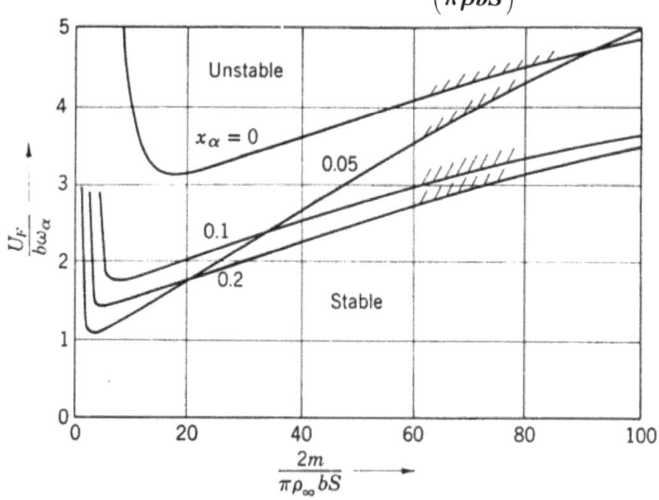

For low values of mass ratio, the flutter boundary follows the expected Pines variation with x_α. As the mass ratio increases, the Pines criteria breaks down and the flutter boundary increases. This is the region normally occupied by modern high-speed aircraft operating in the Earth's atmosphere. The mass ratios where the flutter minimums occur, however, deserve some attention because these are representative values for ultralight and small general aviation airplanes. This suggests that such aircraft may be more susceptible to flutter problems. Finally, note that as the mass ratio goes to zero, the flutter boundaries go to infinity. This implies that human-powered aircraft or underwater wings may be free from flutter considerations.

6.3.6 An Energy Interpretation of Flutter

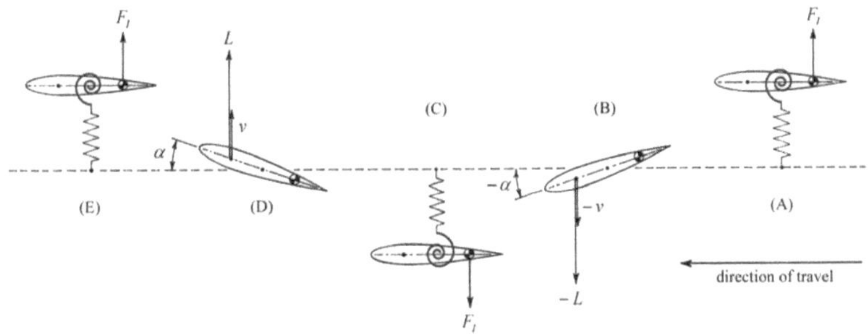

The simple model shown in the figure provides an easy interpretation of flutter. The section is travelling from right to left, and station (A) is the starting point where the section is just beginning its downward plunge. This downward acceleration gives rise to an inertial reaction force at the mass center so that the aft portion of the section lags the downward motion. By the time the section is at station (B), at maximum plunging velocity, it has a negative pitch angle. If this angle is large enough, the relative angle of attack is negative (recall that the relative angle of attack depends on both α and $\dot{h}$), and a negative lift force is generated. The section then moves to station (C), where it then has an upward acceleration, and by the time the section reaches station (D), a positive lift is achieved in the same direction as the positive vertical velocity. The point is that the aerodynamic forces are putting energy into the system because lift is acting in the vertical velocity direction, both on the downstroke and the upstroke. If this input energy matches the energy absorbed by damping, then the flutter condition is achieved.

It should also be observed how the aft location of the mass center drives the phase angle between plunging and pitching such that flutter is produced. By contrast, if the mass center is forward of the elastic axis, the previous model can be redrawn to show how energy is taken out of the system and flutter is suppressed. This provides a physical interpretation of the Pines criteria. In fact, suspending a mass forward of a wing is a common technique for eliminating flutter. For example, a picture of a B-47, such as discussed in the "Divergence" chapter, is shown to illustrate the forward mounting of the engines for flutter suppression.

6.4 FINITE WING

For the previous "representative-section" analysis, it was assumed that this could be related to a finite wing by choosing specific properties at a certain spanwise location (e.g., 75% semispan). This assumption will be more rigorously explored in the finite-wing analysis in this section. Further, this analysis is mandatory when there are large local variations in inertial, elastic, and geometric properties (such as caused by cutouts, engine mounts, etc.). Such abrupt variations reduce the veracity of the representative-section assumption.

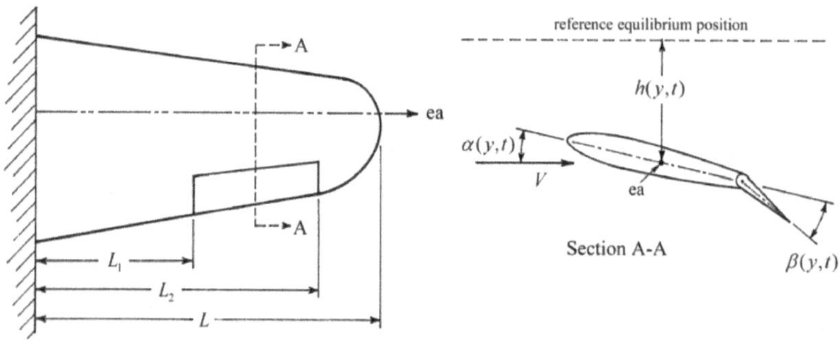

The approach used will be the assumed mode (or Rayleigh-Ritz) analysis. The modes will be provided by the unforced-beam solutions discussed in the "Structural Dynamics" chapter.

6.4.1 Uncoupled Assumed Modes

The assumed modes are obtained from a beam where the elastic axis coincides with the mass centers (note, though, that the modes eventually solved for are *not* uncoupled):

$$h(y,t) = H(y)q_h(t),\ \alpha(y,t) = A(y)q_\alpha(t),\ \beta(y,t) = B(y)q_\beta(t) \tag{6.23}$$

where
$H(y)$ = the fundamental uncoupled bending mode
$A(y)$ = the fundamental uncoupled twisting mode
$B(y)$ = the aileron mode ($B(y) = const.$ between L_1 and L_2 and equals zero otherwise)

The kinetic energy of a spanwise differential element is

$$dT = \left[\frac{1}{2}M(y)\dot{h}^2 + \frac{1}{2}I_\alpha(y)\dot{\alpha}^2 + \frac{1}{2}I_\beta(y)\dot{\beta}^2 + S_\alpha(y)\dot{h}\dot{\alpha} + S_\beta(y)\dot{h}\dot{\beta} + P_{\alpha\beta}(y)\dot{\alpha}\dot{\beta}\right]dy \tag{6.24}$$

The terms M, I_α, etc. were previously encountered for the representative section. In this case, $M(y)$, $I_\alpha(y)$, etc., are the same terms as applied to a section at a particular y location on the wing. For example:

$$M(y) = m_w(y) + m_\beta(y)$$

where

$$m_w(y) = \text{ wing's mass/span, } m_\beta(y) = \text{ aileron mass/span}$$

When Equations (6.23) are substituted into (6.24) and integrated from 0 to L, one obtains

$$T = \frac{1}{2}\dot{q}_h\ \dot{q}_\alpha\ \dot{q}_\beta\left[\tilde{M}\right]\begin{Bmatrix}\dot{q}_h\\ \dot{q}_\alpha\\ \dot{q}_\beta\end{Bmatrix} \tag{6.25}$$

where

$$\left[\tilde{M}\right] = \begin{bmatrix}\bar{M} & \bar{S}_\alpha & \bar{S}_\beta\\ \bar{S}_\alpha & \bar{I}_\alpha & \bar{P}_{\alpha\beta}\\ \bar{S}_\beta & \bar{P}_{\alpha\beta} & \bar{I}_\beta\end{bmatrix}$$

$$\bar{M} = \int_0^L M(y)\left[H(y)\right]^2 dy,\ \bar{S}_\alpha = \int_0^L S_\alpha(y)\left[H(y)A(y)\right]dy$$

$$\bar{I}_\alpha = \int_0^L I_\alpha(y)\left[A(y)\right]^2 dy,\ \bar{S}_\beta = \int_{L_1}^{L_2} S_\beta(y)\left[H(y)B(y)\right]dy$$

$$\bar{I}_\beta = \int_{L_1}^{L_2} I_\beta(y)\left[B(y)\right]^2 dy,\ \bar{P}_{\alpha\beta} = \int_{L_1}^{L_2} P_{\alpha\beta}(y)\left[A(y)B(y)\right]dy$$

The elastic energy may be expressed as

$$V = V_{bending} + V_{twisting} + V_{aileron}$$

where

$$V_{bending} = \frac{1}{2}\int_0^L EI\left(h''\right)^2 dy = \left\{\frac{1}{2}\int_0^L EI\left[H''(y)\right]^2 dy\right\}q_h^2 \equiv \frac{1}{2}\bar{K}_h q_h^2 \tag{6.26}$$

$$V_{twisting} = \frac{1}{2}\int_0^L GJ\left(\alpha'\right)^2 dy = \left\{\frac{1}{2}\int_0^L GJ\left[A'(y)\right]^2 dy\right\}q_\alpha^2 \equiv \frac{1}{2}\bar{K}_\alpha q_\alpha^2 \tag{6.27}$$

Also, the aileron is assumed to act against a torsional spring of stiffness K_β so that one obtains

$$V_{aileron} = \frac{1}{2}\int_{L_1}^{L_2} K_\beta \beta^2 dy = \left\{\frac{1}{2}\int_{L_1}^{L_2} K_\beta\left[B(y)\right]^2 dy\right\}q_\beta^2 \equiv \frac{1}{2}\bar{K}_\beta q_\beta^2 \tag{6.28}$$

Equations (6.26) through (6.28) combine to give

$$V = \frac{1}{2}\left\lfloor q_h\ q_\alpha\ q_\beta\right\rfloor\left[\tilde{K}\right]\begin{Bmatrix}q_h\\ q_\alpha\\ q_\beta\end{Bmatrix} \tag{6.29}$$

where

$$[\tilde{K}] = \begin{bmatrix} \bar{K}_h & 0 & 0 \\ 0 & \bar{K}_\alpha & 0 \\ 0 & 0 & \bar{K}_\beta \end{bmatrix}$$

If the mass centers and elastic axes all coincided, the bending, twisting, and aileron vibrations would be uncoupled, and the unforced dynamic equations would be

$$\bar{M}\ddot{h} + \bar{K}_h h = 0,\ \bar{I}_\alpha\ddot{\alpha} + \bar{K}_\alpha\alpha = 0,\ \bar{I}_\beta\ddot{\beta} + \bar{K}_\beta\beta = 0$$

One may then obtain the natural frequencies

$$\omega_h^2 = \frac{\bar{K}_h}{\bar{M}},\ \omega_\alpha^2 = \frac{\bar{K}_\alpha}{\bar{I}_\alpha},\ \omega_\beta^2 = \frac{\bar{K}_\beta}{\bar{I}_\beta}$$

and these allow the elastic matrix to be written as

$$[\tilde{K}] = \begin{bmatrix} \bar{M}\omega_h^2 & 0 & 0 \\ 0 & \bar{I}_\alpha\omega_\alpha^2 & 0 \\ 0 & 0 & \bar{I}_\beta\omega_\beta^2 \end{bmatrix}$$

For the structural damping, one may draw upon the development for the representative section to define a "dissipation function":

$$D = \frac{1}{2}\lfloor \dot{q}_h\ \dot{q}_\alpha\ \dot{q}_\beta \rfloor [\tilde{D}] \begin{Bmatrix} \dot{q}_h \\ \dot{q}_\alpha \\ \dot{q}_\beta \end{Bmatrix} \tag{6.30}$$

where

$$[\tilde{D}] = \begin{bmatrix} \bar{K}_h g_h/\omega & 0 & 0 \\ 0 & \bar{K}_\alpha g_\alpha/\omega & 0 \\ 0 & 0 & \bar{K}_\beta g_\beta/\omega \end{bmatrix} = \begin{bmatrix} \bar{M}g_h\omega_h^2/\omega & 0 & 0 \\ 0 & \bar{I}_\alpha g_\alpha\omega_\alpha^2/\omega & 0 \\ 0 & 0 & \bar{I}_\beta g_\beta\omega_\beta^2/\omega \end{bmatrix}$$

and the g values are the structural damping coefficients.

The perturbation aerodynamic force and moments are distributed along the span:

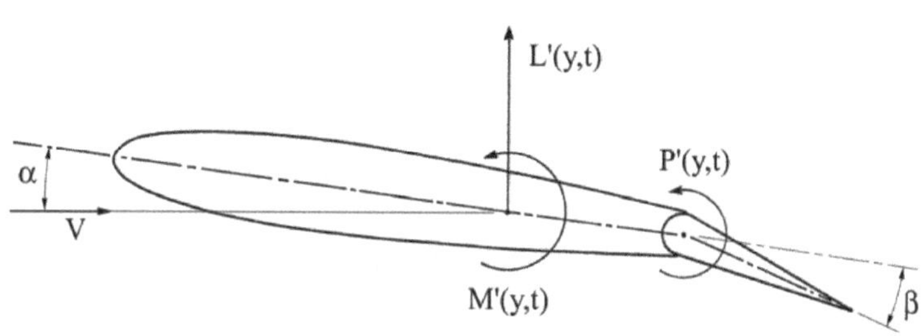

$$L'(y,t)=\frac{Lift}{Span},\ M'(y,t)=\frac{Moment}{Span},\ P'(y,t)=\frac{Aileron\ Hinge\ Moment}{Span}$$

Therefore, the generalized force matrix is

$$\{F'\}=\begin{Bmatrix}-\bar{L}'\\-\bar{M}'\\-\bar{P}'\end{Bmatrix}$$

where

$$\bar{L}'=\int_0^L H(y)L'(y,t)dy,\ \bar{M}'=\int_0^L A(y)M'(y,t)dy,\ \bar{P}'=\int_{L_1}^{L_2} B(y)P'(y,t)dy$$

Now, when Equations (6.25), (6.29), and (6.30) are substituted into Lagrange's equation, one obtains

$$[\tilde{M}]\begin{Bmatrix}\ddot{q}_h\\\ddot{q}_\alpha\\\ddot{q}_\beta\end{Bmatrix}+[\tilde{D}]\begin{Bmatrix}\dot{q}_h\\\dot{q}_\alpha\\\dot{q}_\beta\end{Bmatrix}+[\tilde{K}]\begin{Bmatrix}q_h\\q_\alpha\\q_\beta\end{Bmatrix}=\begin{Bmatrix}-\bar{L}'\\-\bar{M}'\\-\bar{P}'\end{Bmatrix} \tag{6.31}$$

Observe the similarity of these equations to those for the representative section. Differences are that q_h, q_α, q_β replace h, α, β, and the matrix terms are now integral quantities. This gave rise to the notion that the finite-wing case may be dynamically approximated by a representative section at some point along the semi-span (usually at $\approx 0.75L$). This notion will be demonstrated for the forthcoming numerical example.

The strip-theory assumption is applied to this finite-wing model. That is, the two-dimensional aerodynamic equations are assumed to apply to each section along the span. For example, the lift force for the *fixed aileron* case may be expressed (from Equation (6.18)) as

$$L'(y,t)=\rho V^2 A'_{hh}(y)h(y,t)+\rho V^2 A'_{h\alpha}(y)\alpha(y,t)+\rho V A''_{hh}(y)\dot{h}(y,t)+\rho V A''_{h\alpha}(y)\dot{\alpha}(y,t)$$

So the generalized lift force is obtained from

$$\begin{aligned}\bar{L}'=\int_0^L H(y)L'(y,t)dy&=\left\{\rho V^2\int_0^L A'_{hh}(y)[H(y)]^2\,dy\right\}q_h\\&+\left\{\rho V^2\int_0^L A'_{h\alpha}(y)H(y)A(y)dy\right\}q_\alpha+\left\{\rho V\int_0^L A''_{hh}(y)[H(y)]^2\,dy\right\}\dot{q}_h\\&+\left\{\rho V\int_0^L A''_{h\alpha}(y)H(y)A(y)dy\right\}\dot{q}_\alpha\end{aligned} \tag{6.32}$$

This can be extended to the free aileron case and likewise applied to $M'(y,t)$ and $P'(y,t)$ so that the generalized force matrix may be represented as

$$\begin{Bmatrix} -\bar{L}' \\ -\bar{M}' \\ -\bar{P}' \end{Bmatrix} = -\rho V^2 \left[\bar{A}_S\right] \begin{Bmatrix} q_h \\ q_\alpha \\ q_\beta \end{Bmatrix} - \rho V \left[\bar{A}_D\right] \begin{Bmatrix} \dot{q}_h \\ \dot{q}_\alpha \\ \dot{q}_\beta \end{Bmatrix}$$

where, as before, $\left[\bar{A}_S\right]$ may be interpreted as an aerodynamic "spring" matrix and $\left[\bar{A}_D\right]$ may be interpreted as an aerodynamic "damping" matrix. Both of these matrices are square and composed of integral terms, such as in Equation (6.32).

It should be noted that for wing planforms where the chord, $2b$, is varying along the span, k will also vary along the span:

$$k(y) = \omega b(y) / V$$

Usually, the value of b at $0.75L$ is chosen for calculating k.

6.4.2 Discrete-Element Numerical Example

Martin Goland obtained the exact solution for a rectangular-planform wing of semispan $L = 6.096\ m$ and half-chord $b = 0.9144\ m$ ("The Flutter of a Uniform Cantilever Wing," *Journal of Applied Mechanics*, vol. 12, no. 4, December 1945):

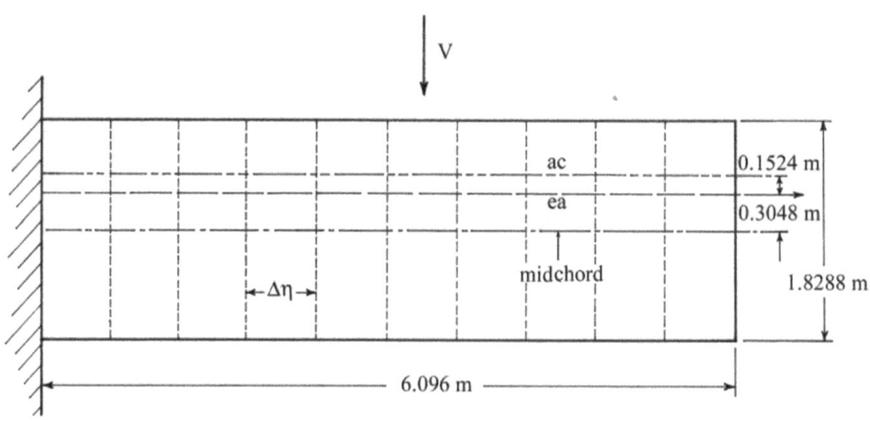

The spanwise properties are constant:

$$M(y) = 35.7087\ kg/m;\ I_\alpha(y) = 8.6406\ kg - m^2/m;\ S_\alpha(y) = 6.5217\ kg - m/m$$
$$EI(y) = 9.7523 \times 10^6\ N - m^2;\ GJ(y) = 9.8763 \times 10^5\ N - m^2/rad$$

Also,

$$a(y) = -\frac{1}{3},\ \text{so that } ba = -0.3048m \text{ and } bx = e(y) = 0.1524m$$

If the wing is divided into ten equal spanwise segments of width $\Delta\eta = 0.6096\ m$, then the discrete-element values are

$$M_i = M(y)\Delta\eta = 21.7680\ kg; (I_\alpha)_i = I_\alpha(y)\Delta\eta = 5.2673\ kg - m^2$$

$$(S_\alpha)_i = S_\alpha(y)\Delta\eta = 3.9756\ kg - m,\ i = 1{,}10$$

Upon using the methods described in Section 5.3.3, the uncoupled dynamic modes may be found. The first bending and torsional modes are

$$\{H\} = \begin{Bmatrix} 0.0027 \\ 0.0234 \\ 0.0617 \\ 0.1147 \\ 0.1795 \\ 0.2531 \\ 0.3330 \\ 0.4170 \\ 0.5031 \\ 0.5901 \end{Bmatrix}, \{A\} = \begin{Bmatrix} 0.0351 \\ 0.1044 \\ 0.1711 \\ 0.2337 \\ 0.2904 \\ 0.3401 \\ 0.3813 \\ 0.4132 \\ 0.4349 \\ 0.4458 \end{Bmatrix}$$

with uncoupled frequencies of

$$\omega_h = 49.562\ rad/\sec,\ \omega_\alpha = 87.039\ rad/\sec$$

One may now obtain the terms in the $\left[\tilde{M}\right]$ matrix:

$$\bar{M} = \int_0^L M(y)\left[H(y)\right]^2 dy \approx \lfloor M \rfloor \left\{H^2\right\} = 21.7659$$

$$\bar{S}_\alpha = \int_0^L S_\alpha(y)\left[H(y)A(y)\right] dy \approx \lfloor S_\alpha \rfloor \left\{HA\right\} = 3.8127$$

$$\bar{I}_\alpha = \int_0^L I_\alpha(y)\left[A(y)\right]^2 dy \approx \lfloor I_\alpha \rfloor \left\{A^2\right\} = 5.2674$$

And the terms in the $\left[\tilde{K}\right]$ matrix follow:

$$\bar{K}_h = \bar{M}\omega_h^2 = 5.3466\times10^4,\ \bar{K}_\alpha = \bar{I}_\alpha\omega_\alpha^2 = 3.9905\times10^4$$

For the aerodynamic matrix, the terms may be evaluated as shown for $\bar{A}'_{hh}$, $\bar{A}'_{\alpha h}$, and $\bar{A}'_{\alpha\alpha}$:

$$\bar{A}'_{hh} = \int_0^L A'_{hh}(y)\left[H(y)\right]^2 dy \approx \left\lfloor A'_{hh} \right\rfloor \left\{(H)^2 \Delta\eta\right\}$$

$$\bar{A}'_{h\alpha} = \int_0^L A'_{h\alpha}(y)H(y)A(y)dy \approx \lfloor A'_{h\alpha} \rfloor \{HA\Delta\eta\}$$

$$\bar{A}'_{\alpha h} = \int_0^L A'_{\alpha h}(y)A(y)H(y)dy \approx \lfloor A'_{\alpha h} \rfloor \{AH\Delta\eta\}$$

$$\bar{A}'_{\alpha\alpha} = \int_0^L A'_{\alpha\alpha}(y)[A(y)]^2 dy \approx \lfloor A'_{\alpha\alpha} \rfloor \{(A)^2 \Delta\eta\}$$

With a reduced frequency, $k = 0.5215$, and atmospheric density, $\rho = 1.225\ kg/m^3$, the aerodynamic matrix terms are

$$[\bar{A}_S] = \begin{bmatrix} \bar{A}'_{hh} & \bar{A}'_{h\alpha} \\ \bar{A}'_{\alpha h} & \bar{A}'_{\alpha\alpha} \end{bmatrix} = \begin{bmatrix} -0.2257 & 2.0567 \\ -0.1955 & -0.4538 \end{bmatrix}$$

$$[\bar{A}_D] = \begin{bmatrix} \bar{A}''_{hh} & \bar{A}''_{h\alpha} \\ \bar{A}''_{\alpha h} & \bar{A}''_{\alpha\alpha} \end{bmatrix} = \begin{bmatrix} 2.0776 & 2.1842 \\ -0.3038 & 1.1172 \end{bmatrix}$$

Structural damping will be ignored, so upon defining

$$\{q\} \equiv \begin{Bmatrix} q_h \\ q_\alpha \end{Bmatrix}$$

the dynamic equations become

$$[\tilde{M}]\{\ddot{q}\} + [\tilde{K}]\{q\} = -\rho V^2 [\bar{A}_S]\{q\} - \rho V [\bar{A}_D]\{\dot{q}\}$$

which may be rewritten as

$$[\tilde{M}]\{\ddot{q}\} + \rho V[\bar{A}_D]\{\dot{q}\} + [\tilde{K} + \rho V^2 \bar{A}_S]\{q\} = 0$$

Upon defining

$$[\hat{K}] \equiv [\tilde{K} + \rho V^2 \bar{A}_S]$$

one obtains

$$\{\ddot{q}\} = -\rho V [\tilde{M}]^{-1} [\bar{A}_D]\{\dot{q}\} - [\tilde{M}]^{-1}[\hat{K}]\{q\}$$

This can be rewritten in matrix form as

$$\begin{Bmatrix} \ddot{q} \\ \dot{q} \end{Bmatrix} = [U_{dyn}] \begin{Bmatrix} \dot{q} \\ q \end{Bmatrix}$$

where

$$[U_{dyn}] = \begin{bmatrix} -\rho V [\tilde{M}]^{-1}[\bar{A}_D] & -[\tilde{M}]^{-1}[\hat{K}] \\ [I] & [0] \end{bmatrix}$$

Upon assuming a solution of

$$q = \bar{q}e^{\lambda t}$$

the dynamic equations become

$$\begin{Bmatrix} \lambda^2 \bar{q} \\ \lambda \bar{q} \end{Bmatrix} = \left[U_{dyn}\right] \begin{Bmatrix} \lambda \bar{q} \\ \bar{q} \end{Bmatrix} = \lambda \begin{Bmatrix} \lambda \bar{q} \\ \bar{q} \end{Bmatrix} \rightarrow \left[U_{dyn} - I\lambda\right] \begin{Bmatrix} \lambda \bar{q} \\ \bar{q} \end{Bmatrix} = 0$$

This gives the eigenvalue formulation for the solution of the flutter roots, λ, and the corresponding modal vectors, $\{\bar{q}\}$. For this particular example, the flutter speed is 135.0 $m/\sec$, with a corresponding flutter frequency of 70.4 rad/sec.

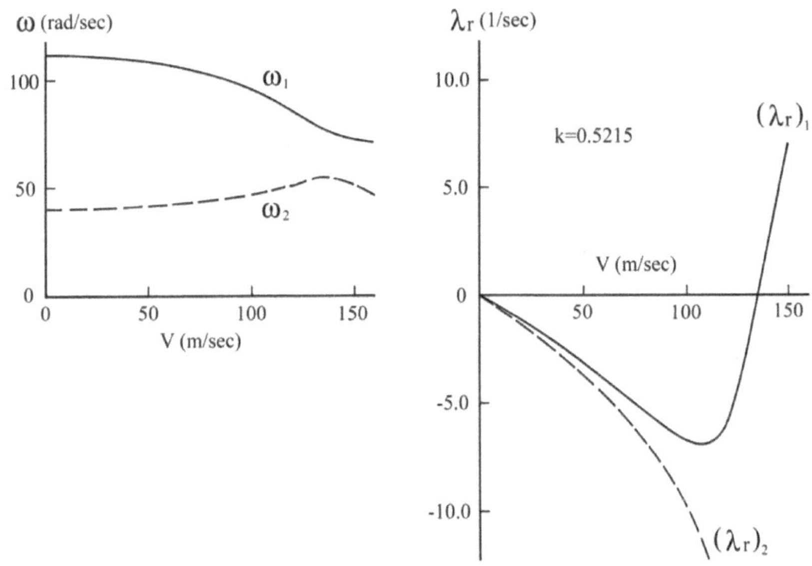

By comparison, Lin and Iliff's updated solution of the Goland example ("Aerodynamic Lift and Moment Calculations Using a Closed-Form Solution of the Possio Equation," NASA/TM-2000-209019) gives the flutter speed as 136 m/sec, with a flutter frequency of 69.9 rad/sec.

One may also calculate the flutter speed using a *representative section* approach, which assumes that M, S_α, and I_α are the running values (values per unit span) at the $0.75L$ location (for this example, these are constant along the span).

$$M = 35.7087\ kg;\ I_\alpha = 8.6406\ kg - m^2;\ S_\alpha = 6.5217\ kg - m$$

One also has that

$$b = 0.9144\ m,\ ba = -0.3048,\ x = a + 0.5 = 0.1667$$

The fundamental frequencies, ω_h and ω_α, are assumed to be those previously obtained for the finite wing:

$$\omega_h = 49.562\ rad/\sec,\ \omega_\alpha = 87.039\ rad/\sec$$

With this information, one may use the methods previously described to obtain

$$V_{flutter} = 127.8\ m/\sec,\ \omega_{flutter} = 71.9\ rad/\sec$$

This is a reasonable, and conservative, match to the finite-wing value. However, this may not be so for a wing with strongly varying properties or concentrated masses along the span. Historically, the representative-section approach was most useful when computing was limited and laborious. Its best use now is for rough approximate calculations (and, of course, academic examples).

6.4.3 Coupled Assumed Modes

A major assumption in the previous analysis is that the flutter modes have the shape of the fundamental unforced uncoupled modes. Although Goland has shown this to be a reasonable assumption for his example, Theodorsen and Garrick have made a strong case for this not being generally so ("Mechanism of Flutter: A Theoretical and Experimental Investigation of the Flutter Problem," NACA TR 685, 1940). For greater accuracy, therefore, multiple assumed modes may be chosen:

$$h(y,t) = \sum_{i=1}^{n} H_i(y) q_i(t),\ \alpha(y,t) = \sum_{i=1}^{n} A_i(y) q_i(t),\ \beta(y,t) = \sum_{i=1}^{n} B_i(y) q_i(t) \quad (6.33)$$

Usually, $n = 3$ is adequate so that one has

$$\begin{Bmatrix} h \\ \alpha \\ \beta \end{Bmatrix} = \begin{bmatrix} H_1 & H_2 & H_3 \\ A_1 & A_2 & A_3 \\ B_1 & B_2 & B_3 \end{bmatrix} \begin{Bmatrix} q_1 \\ q_2 \\ q_3 \end{Bmatrix}$$

The development of the equations is an elaboration on the derivation of those for the fundamental mode case. For example, starting from Equation (6.24), the substitution of the assumed modes gives the kinetic energy of a differential element as

$$dT = \frac{1}{2} M(y)\left(H_1\dot{q}_1 + H_2\dot{q}_2 + H_3\dot{q}_3\right)^2 dy + \ldots.$$
$$+ S_\alpha(y)\left(H_1\dot{q}_1 + H_2\dot{q}_2 + H_3\dot{q}_3\right)\times\left(A_1\dot{q}_1 + A_2\dot{q}_2 + A_3\dot{q}_3\right) dy + \ldots.$$

This expands to give

$$dT = \left[\frac{M(y)}{2}\left(H_1^2\dot{q}_1^2 + 2H_1H_2\dot{q}_1\dot{q}_2 + 2H_1H_3\dot{q}_1\dot{q}_3 + H_2^2\dot{q}_2^2 + 2H_2H_3\dot{q}_2\dot{q}_3 + H_3^2\dot{q}_3^2\right)\right] dy + \ldots.$$
$$+ \left[S_\alpha(y)\left(H_1A_1\dot{q}_1^2 + 2H_1A_2\dot{q}_1\dot{q}_2 + 2H_1A_3\dot{q}_1\dot{q}_3 + H_2A_2\dot{q}_2^2 + 2H_2A_3\dot{q}_2\dot{q}_3 + H_3A_3\dot{q}_3^2\right)\right] dy + ..$$

which may also be expressed as

$$dT = \left[\frac{1}{2} \sum_{i=1}^{3} \sum_{j=1}^{3} M(y) H_i H_j \dot{q}_i \dot{q}_j + \ldots + \sum_{i=1}^{3} \sum_{j=1}^{3} S_\alpha(y) H_i A_j \dot{q}_i \dot{q}_j + \ldots \right] dy$$

When this is integrated between 0 and L, one obtains

$$T = \frac{1}{2} \sum_{i=1}^{3} \sum_{j=1}^{3} \bar{M}_{ij} \dot{q}_i \dot{q}_j + \ldots + \sum_{i=1}^{3} \sum_{j=1}^{3} \left(\bar{S}_\alpha \right)_{ij} \dot{q}_i \dot{q}_j + \ldots \tag{6.34}$$

where

$$\bar{M}_{ij} = \int_0^L M(y) H_i(y) H_j(y) dy \text{ (compare with } M_{zz_{ij}} = \int_0^L m Z_i Z_j \, dy)$$

$$\left(\bar{S}_\alpha \right)_{ij} = \int_0^L S_\alpha(y) H_i(y) A_j(y) dy \text{ (compare with } M_{z\theta_{ij}} = -\int_0^L m \tilde{d} \Theta_i Z_j \, dy)$$

$$\left(\bar{I}_\alpha \right)_{ij} = \int_0^L I_\alpha(y) A_i(y) A_j(y) dy \text{ (compare with } M_{\theta\theta_{ij}} = \int_0^L mR^2 \Theta_i \Theta_j \, dy)$$

$$\left(\bar{S}_\beta \right)_{ij} = \int_{L_1}^{L_2} S_\beta(y) H_i(y) B_j(y) dy, \; \left(\bar{I}_\beta \right)_{ij} = \int_{L_1}^{L_2} I_\beta(y) B_i(y) B_j(y) dy$$

$$\left(\bar{P}_{\alpha\beta} \right)_{ij} = \int_0^L P_{\alpha\beta}(y) A_i(y) B_j(y) dy$$

Observe how this model has certain identical dynamic terms to the coupled bending and twisting case discussed in Section 5.4.2. Now, define

$$\tilde{M}_{ij} \equiv \bar{M}_{ij} + \left(\bar{I}_\alpha \right)_{ij} + \left(\bar{I}_\beta \right)_{ij} + 2\left(\bar{S}_\alpha \right)_{ij} + 2\left(\bar{S}_\beta \right)_{ij} + 2\left(\bar{P}_{\alpha\beta} \right)_{ij}$$

so that Equation (6.34) becomes

$$T = \frac{1}{2} \sum_{i=1}^{3} \sum_{j=1}^{3} \tilde{M}_{ij} \dot{q}_i \dot{q}_j \tag{6.35}$$

In matrix form, this may be written as

$$T = \frac{1}{2} \lfloor \dot{q}_1 \; \dot{q}_2 \; \dot{q}_3 \rfloor \begin{bmatrix} \tilde{M}_{1,1} & \tilde{M}_{1,2} & \tilde{M}_{1,3} \\ \tilde{M}_{2,1} & \tilde{M}_{2,2} & \tilde{M}_{2,3} \\ \tilde{M}_{3,1} & \tilde{M}_{3,2} & \tilde{M}_{3,3} \end{bmatrix} \begin{Bmatrix} \dot{q}_1 \\ \dot{q}_2 \\ \dot{q}_3 \end{Bmatrix} = \frac{1}{2} \lfloor \dot{q} \rfloor \left[\tilde{M} \right] \{ \dot{q} \} \tag{6.36}$$

The bending elastic energy is

$$V_{bending} = \frac{1}{2} \int_0^L EI \left(h'' \right)^2 dy = \frac{1}{2} \int_0^L EI \left(H_1'' q_1 + H_2'' q_2 + H_3'' q_3 \right)^2 dy$$

which becomes

$$V_{bending}=\frac{1}{2}\left[\begin{aligned}&\left(\bar{K}_h\right)_{1,1}q_1^2+\left(\bar{K}_h\right)_{2,2}q_2^2+\left(\bar{K}_h\right)_{3,3}q_3^2\\&+2\left(\bar{K}_h\right)_{1,2}q_1q_2+2\left(\bar{K}_h\right)_{1,3}q_1q_3+2\left(\bar{K}_h\right)_{2,3}q_2q_3\end{aligned}\right]$$

where

$$\left(\bar{K}_h\right)_{i,j}=\int_0^L EI\,H_i''H_j''dy$$

(Compare this with $K_{z_{ij}}=\int_0^L EI\,Z_i''Z_j''dy$ from the coupled bending and twisting case.)

Likewise, the twisting strain energy is given by

$$V_{twisting}=\frac{1}{2}\int_0^L GJ\left(\alpha'\right)^2dy=\frac{1}{2}\left[\begin{aligned}&\left(\bar{K}_\alpha\right)_{1,1}q_1^2+\left(\bar{K}_\alpha\right)_{2,2}q_2^2+\left(\bar{K}_\alpha\right)_{3,3}q_3^2\\&+2\left(\bar{K}_\alpha\right)_{1,2}q_1q_2+2\left(\bar{K}_\alpha\right)_{1,3}q_1q_3+2\left(\bar{K}_\alpha\right)_{2,3}q_2q_3\end{aligned}\right]$$

where

$$\left(\bar{K}_\alpha\right)_{i,j}=\int_0^L GJ\,A_i'A_j'dy$$

(Compare this with $K_{\theta_{ij}}=\int_0^L GJ\,\Theta_i'\Theta_j'\,dy$ from the coupled bending and twisting case.)

If the aileron acts against a distributed spring, $K_\beta(y)$, then the strain energy is given by

$$V_{aileron}=\frac{1}{2}\int_{L_1}^{L_2}K_\beta\beta^2dy=\frac{1}{2}\left[\begin{aligned}&\left(\bar{K}_\beta\right)_{1,1}q_1^2+\left(\bar{K}_\beta\right)_{2,2}q_2^2+\left(\bar{K}_\beta\right)_{3,3}q_3^2\\&+2\left(\bar{K}_\beta\right)_{1,2}q_1q_2+2\left(\bar{K}_\beta\right)_{1,3}q_1q_3+2\left(\bar{K}_\beta\right)_{2,3}q_2q_3\end{aligned}\right]$$

where

$$\left(\bar{K}_\beta\right)_{i,j}=\int_{L_1}^{L_2}K_\beta\,B_iB_jdy$$

Upon defining

$$\tilde{K}_{i,j}\equiv\left(\bar{K}_h\right)_{i,j}+\left(\bar{K}_\alpha\right)_{i,j}+\left(\bar{K}_\beta\right)_{i,j}$$

one obtains that

$$V=V_{bending}+V_{twisting}+V_{aileron}=\frac{1}{2}\sum_{i=1}^{3}\sum_{j=1}^{3}\tilde{K}_{ij}q_iq_j \tag{6.37}$$

which, in matrix form, is

$$V = \frac{1}{2}\lfloor q_1\, q_2\, q_3 \rfloor \begin{bmatrix} \tilde{K}_{1,1} & \tilde{K}_{1,2} & \tilde{K}_{1,3} \\ \tilde{K}_{2,1} & \tilde{K}_{2,2} & \tilde{K}_{2,3} \\ \tilde{K}_{3,1} & \tilde{K}_{3,2} & \tilde{K}_{3,3} \end{bmatrix} \begin{Bmatrix} q_1 \\ q_2 \\ q_3 \end{Bmatrix} = \frac{1}{2} q\left[\tilde{K}\right]\{q\} \tag{6.38}$$

Note that if *coupled* normal modes are chosen, then the inertial matrix in Equation (6.36) becomes

$$\left[\tilde{M}\right] = \left[\tilde{M}_{diag.}\right] = \begin{bmatrix} \tilde{M}_1 & 0 & 0 \\ 0 & \tilde{M}_2 & 0 \\ 0 & 0 & \tilde{M}_3 \end{bmatrix}$$

where

$$\tilde{M}_i = \int_0^L \left\{ M(y)\left[H_i(y)\right]^2 + I_\alpha(y)\left[A_i(y)\right]^2 + I_\beta(y)\left[B_i(y)\right]^2 \right\} dy$$
$$+2\int_0^L \left\{ S_\alpha(y) H_i(y) A_i(y) + S_\beta(y) H_i(y) B_i(y) + P_{\alpha\beta}(y) A_i(y) B_i(y) \right\} dy$$

Likewise, the stiffness matrix becomes

$$\left[\tilde{K}\right] = \left[\tilde{K}_{diag.}\right] = \begin{bmatrix} \tilde{K}_1 & 0 & 0 \\ 0 & \tilde{K}_2 & 0 \\ 0 & 0 & \tilde{K}_3 \end{bmatrix}$$

where

$$\tilde{K}_i = \left(\bar{K}_h\right)_{i,i} + \left(\bar{K}_\alpha\right)_{i,i} + \left(\bar{K}_\beta\right)_{i,i}$$

The off-diagonal terms are zero because of the orthogonality of the normal modes. Upon substituting these $\left[\tilde{M}_{diag.}\right]$ and $\left[\tilde{K}_{diag.}\right]$ matrices into Equations (6.36) and (6.38) and applying Lagrange's equation, one may obtain the approximate (assumed mode) dynamic equations for the unforced undamped case:

$$\begin{bmatrix} \tilde{M}_1 & 0 & 0 \\ 0 & \tilde{M}_2 & 0 \\ 0 & 0 & \tilde{M}_3 \end{bmatrix} \begin{Bmatrix} \ddot{q}_1 \\ \ddot{q}_2 \\ \ddot{q}_3 \end{Bmatrix} + \begin{bmatrix} \tilde{K}_1 & 0 & 0 \\ 0 & \tilde{K}_2 & 0 \\ 0 & 0 & \tilde{K}_3 \end{bmatrix} \begin{Bmatrix} q_1 \\ q_2 \\ q_3 \end{Bmatrix} = 0$$

And the modal frequencies are readily obtained as

$$\omega_1^2 = \frac{\tilde{K}_1}{\tilde{M}_1},\ \omega_2^2 = \frac{\tilde{K}_2}{\tilde{M}_2},\ \omega_3^2 = \frac{\tilde{K}_3}{\tilde{M}_3}$$

Therefore, $\left[\tilde{K}_{diag.}\right]$ may be rewritten as

$$\left[\tilde{K}_{diag.}\right]=\begin{bmatrix}\tilde{M}_1\omega_1^2 & 0 & 0\\ 0 & \tilde{M}_2\omega_2^2 & 0\\ 0 & 0 & \tilde{M}_3\omega_3^2\end{bmatrix}$$

This shows the usefulness of the normal modes, in that the calculation of $\tilde{K}_i$ by using the frequency terms (even obtained experimentally) can be more accurate than finding $\tilde{K}_i$ from derivatives of curve-fitted equations to the mode shapes. Also, the normal-mode formulation gives a structural dissipation function expression in complete analogy to that from the previous section:

$$D=\frac{1}{2}\dot{q}_1\ \dot{q}_2\ \dot{q}_3\left[\tilde{D}\right]\begin{Bmatrix}\dot{q}_1\\ \dot{q}_2\\ \dot{q}_3\end{Bmatrix} \tag{6.39}$$

where

$$\left[\tilde{D}\right]=\begin{bmatrix}\tilde{K}_1g_1/\omega & 0 & 0\\ 0 & \tilde{K}_2g_2/\omega & 0\\ 0 & 0 & \tilde{K}_3g_3/\omega\end{bmatrix}=\begin{bmatrix}\tilde{M}_1g_1\omega_1^2/\omega & 0 & 0\\ 0 & \tilde{M}_2g_2\omega_2^2/\omega & 0\\ 0 & 0 & \tilde{M}_3g_3\omega_3^2/\omega\end{bmatrix}$$

When Lagrange's equation is applied to the inertial, elastic, and dissipation-energy expressions, one obtains

$$\begin{bmatrix}\tilde{M}_1 & 0 & 0\\ 0 & \tilde{M}_2 & 0\\ 0 & 0 & \tilde{M}_3\end{bmatrix}\begin{Bmatrix}\ddot{q}_1\\ \ddot{q}_2\\ \ddot{q}_3\end{Bmatrix}+\begin{bmatrix}\tilde{M}_1g_1\omega_1^2/\omega & 0 & 0\\ 0 & \tilde{M}_2g_2\omega_2^2/\omega & 0\\ 0 & 0 & \tilde{M}_3g_3\omega_3^2/\omega\end{bmatrix}\begin{Bmatrix}\dot{q}_1\\ \dot{q}_2\\ \dot{q}_3\end{Bmatrix}$$
$$+\begin{bmatrix}\tilde{M}_1\omega_1^2 & 0 & 0\\ 0 & \tilde{M}_2\omega_2^2 & 0\\ 0 & 0 & \tilde{M}_3\omega_3^2\end{bmatrix}\begin{Bmatrix}q_1\\ q_2\\ q_3\end{Bmatrix}=\begin{Bmatrix}\tilde{Q}_1\\ \tilde{Q}_2\\ \tilde{Q}_3\end{Bmatrix}$$

The generalized force matrix, $\left\{\tilde{Q}\right\}$, comes from the aerodynamic force and moments. Also, strip theory is assumed as in the previous section. Therefore, the lift expression is

$$L'(y,t)=\rho V^2A'_{hh}(y)h(y,t)+\rho V^2A'_{h\alpha}(y)\alpha(y,t)+\rho V^2A'_{h\beta}(y)\beta(y,t)$$
$$+\rho VA''_{hh}(y)\dot{h}(y,t)+\rho VA''_{h\alpha}(y)\dot{\alpha}(y,t)+\rho VA''_{h\beta}(y)\dot{\beta}(y,t)$$

With the assumed modes, this becomes

$$
\begin{aligned}
L' = {} & \rho V^2 A'_{hh}(H_1 q_1 + H_2 q_2 + H_3 q_3) + \rho V^2 A'_{h\alpha}(A_1 q_1 + A_2 q_2 + A_3 q_3) \\
& + \rho V^2 A'_{h\beta}(B_1 q_1 + B_2 q_2 + B_3 q_3) + \rho V^2 A''_{hh}(H_1 \dot{q}_1 + H_2 \dot{q}_2 + H_3 \dot{q}_3) \\
& + \rho V A''_{h\alpha}(A_1 \dot{q}_1 + A_2 \dot{q}_2 + A_3 \dot{q}_3) + \rho V^2 A''_{h\beta}(B_1 \dot{q}_1 + B_2 \dot{q}_2 + B_3 \dot{q}_3)
\end{aligned}
$$

Note that Equations (6.33) are of the form

$$z_i = \sum_{i=1}^{n} Q_{ij} q_j \rightarrow \dot{z}_i = \sum_{i=1}^{n} Q_{ij} \dot{q}_j$$

Therefore, applied forces, F_j, are transformed into generalized forces, $\tilde{Q}_j$, by

$$\tilde{Q}_j = \sum_{i=1}^{n} Q_{ij} F_i \rightarrow d\tilde{Q}_j = \sum_{i=1}^{n} Q_{ij}\, dF_i$$

In this case,

$$dF_1 = -L'dy,\ dF_2 = -M'dy,\ dF_3 = -P'dy$$

and

$$Q_{1,j} = H_j,\ Q_{2,j} = A_j,\ Q_{3,j} = B_j$$

so that

$$d\tilde{Q}_1 = -\left(H_1 L' + A_1 M' + B_1 P'\right) dy$$

$$d\tilde{Q}_2 = -\left(H_2 L' + A_2 M' + B_2 P'\right) dy$$

$$d\tilde{Q}_3 = -\left(H_3 L' + A_3 M' + B_3 P'\right) dy$$

Upon substituting the expressions for L', M', and P' and integrating, one obtains the generalized aerodynamic force and moments:

$$
\begin{aligned}
\tilde{Q}_1 = {} & -\rho V^2 \left[(A_S)_{1,1} q_1 + (A_S)_{1,2} q_2 + (A_S)_{1,3} q_3 \right] \\
& -\rho V \left[(A_D)_{1,1} \dot{q}_1 + (A_D)_{1,2} \dot{q}_2 + (A_D)_{1,3} \dot{q}_3 \right]
\end{aligned}
$$

$$
\begin{aligned}
\tilde{Q}_2 = {} & -\rho V^2 \left[(A_S)_{2,1} q_1 + (A_S)_{2,2} q_2 + (A_S)_{2,3} q_3 \right] \\
& -\rho V \left[(A_D)_{2,1} \dot{q}_1 + (A_D)_{2,2} \dot{q}_2 + (A_D)_{2,3} \dot{q}_3 \right]
\end{aligned}
$$

$$\tilde{Q}_3 = -\rho V^2 \left[(A_S)_{3,1} q_1 + (A_S)_{3,2} q_2 + (A_S)_{3,3} q_3 \right] - \rho V \left[(A_D)_{3,1} \dot{q}_1 + (A_D)_{3,2} \dot{q}_2 + (A_D)_{3,3} \dot{q}_3 \right]$$

where, for example,

$$(A_S)_{1,1} = \int_0^L \left(A'_{hh} H_1^2 + A'_{h\alpha} A_1 H_1 + A'_{\alpha h} H_1 A_1 + A'_{\alpha\alpha} A_1^2 \right) dy + \int_{L_1}^{L_2} \left(A'_{h\beta} B_1 H_1 + A'_{\alpha\beta} B_1 A_1 + A'_{\beta h} H_1 B_1 + A'_{\beta\alpha} A_1 B_1 + A'_{\beta\beta} B_1^2 \right) dy$$

The equations may be expressed in matrix form as

$$\begin{Bmatrix} \tilde{Q}_1 \\ \tilde{Q}_2 \\ \tilde{Q}_3 \end{Bmatrix} = -\rho V^2 \begin{bmatrix} (A_S)_{1,1} & (A_S)_{1,2} & (A_S)_{1,3} \\ (A_S)_{2,1} & (A_S)_{2,2} & (A_S)_{2,3} \\ (A_S)_{3,1} & (A_S)_{3,2} & (A_S)_{3,3} \end{bmatrix} \begin{Bmatrix} q_1 \\ q_2 \\ q_3 \end{Bmatrix} - \rho V \begin{bmatrix} (A_D)_{1,1} & (A_D)_{1,2} & (A_D)_{1,3} \\ (A_D)_{2,1} & (A_D)_{2,2} & (A_D)_{2,3} \\ (A_D)_{3,1} & (A_D)_{3,2} & (A_D)_{3,3} \end{bmatrix} \begin{Bmatrix} \dot{q}_1 \\ \dot{q}_2 \\ \dot{q}_3 \end{Bmatrix}$$

$$\equiv -\rho V^2 \left[\bar{A}_S \right] \begin{Bmatrix} q_1 \\ q_2 \\ q_3 \end{Bmatrix} - \rho V \left[\bar{A}_D \right] \begin{Bmatrix} \dot{q}_1 \\ \dot{q}_2 \\ \dot{q}_3 \end{Bmatrix}$$

The complete equations are thus:

$$\left[\tilde{M}_{diag.} \right] \begin{Bmatrix} \ddot{q}_1 \\ \ddot{q}_2 \\ \ddot{q}_3 \end{Bmatrix} + \left[\tilde{D} + \rho V \bar{A}_D \right] \begin{Bmatrix} \dot{q}_1 \\ \dot{q}_2 \\ \dot{q}_3 \end{Bmatrix} + \left[\tilde{K}_{diag.} + \rho V^2 \bar{A}_S \right] \begin{Bmatrix} q_1 \\ q_2 \\ q_3 \end{Bmatrix} = 0$$

These equations bear a similarity to those for the fundamental assumed modes in the previous section. In fact, the same solution methodology may be applied. However, the evaluation of the terms in the square matrices is considerably more complex.

6.4.4 Quasi-Normal Modes

The simplest form of the multi-mode model is that for the uncoupled fundamental modes discussed earlier. Note that this is obtained from the previous case if

$$H_2, H_3, A_1, A_3, B_1, B_2 = 0$$

Therefore,

$$q_1 \to q_h, q_2 \to q_\alpha, q_3 \to q_\beta$$

Scanlan and Rosenbaum (*Introduction to the Study of Aircraft Vibration and Flutter*, Macmillan Co., 1962, pp. 261–263) introduced the concept of "quasi-normal" modes, which is intermediate in complexity between the uncoupled and fully coupled cases. In this instance, it is assumed that the aileron modes are uncoupled from the rest of the wing. This is, in fact, a reasonable assumption if the aileron is inertially balanced about its hinge line. Further, in most cases, the aileron is adequately represented by a single mode. Therefore, the assumed modes become

$$h(y,t) = H_1(y)q_1(t) + H_2(y)q_2(t),\ \alpha(y,t) = A_1(y)q_1(t) + A_2(y)q_2(t)$$

$$\beta(y,t) = B_3(y)q_3(t)$$

where the H_i and A_i modes are those coupled modes obtained if the aileron is assumed to be rigidly locked. If these are coupled *normal* modes, then H_i and A_i are orthogonal to each other and terms such as

$$\left(\bar{S}_\alpha\right)_{1,2} = \int_0^L S_\alpha(y) H_1(y) A_2(y)\, dy = 0$$

However, this is not the case for the assumed aileron mode, B_3, which is not coupled to H_i and A_i (normal or otherwise). Therefore, terms such as

$$\left(\bar{S}_\beta\right)_{1,3} = \int_{L_1}^{L_2} S_\beta(y) H_1(y) B_3(y)\, dy \neq 0$$

Thus, the $\left[\tilde{M}\right]$ matrix is, in general, non-diagonal. The $\left[\tilde{K}\right]$ matrix, though, is still diagonal, and one may still relate its terms to the fundamental vibration frequencies, ω_i, of the quasi-normal modes, where ω_1 and ω_2 are obtained from the dynamic equations for the coupled normal modes:

$$\begin{bmatrix} \tilde{M}_1 & 0 \\ 0 & \tilde{M}_2 \end{bmatrix} \begin{Bmatrix} \ddot{q}_1 \\ \ddot{q}_2 \end{Bmatrix} + \begin{bmatrix} \tilde{K}_1 & 0 \\ 0 & \tilde{K}_2 \end{bmatrix} \begin{Bmatrix} q_1 \\ q_2 \end{Bmatrix} = 0$$

And ω_3 for the aileron is obtained from

$$\tilde{M}_3 \ddot{q}_3 + \tilde{K}_3 q_3 = 0$$

Therefore, the terms in the $\left[\tilde{K}\right]$ matrix are

$$\tilde{K}_1 = \tilde{M}_1 \omega_1^2, \tilde{K}_2 = \tilde{M}_2 \omega_2^2, \tilde{K}_3 = \tilde{M}_3 \omega_3^2$$

Likewise, the dissipation function is the same as given in Equation (6.39).

Note that the inertial properties of the aileron are not included in $\tilde{M}_1$ and $\tilde{M}_2$. If these modes are obtained from experiment, it may not be possible to detach the ailerons, and one has to assume that their contribution to the wing's overall inertial properties is small. Another observation is that the equations for the quasi-normal modes assume vibration in a vacuum. In order to better relate these to experimentally determined modes, Scanlan and Rosenbaum (pp. 240–242) proposed that the aerodynamic apparent-mass terms be included. One may analytically interpret these as additions to the inertial terms, modifying the dynamic equations.

Finally, in most cases, the aileron may be modeled as a non-twisting member so that $B_3(y) = B_3$, which is a constant value over its span (as for the uncoupled fundamental-mode case). Further, ailerons are normally freely hinged, with a single constraint provided by the actuation link. Therefore, the notion introduced earlier of a "distributed spring," $K_\beta(y)$, may be simplified to a single constant value:

$$K_\beta = \int_{L_1}^{L_2} K_\beta(y)\,dy$$

Note that for a mechanical cable-actuated control system, such as used for general aviation and ultralight aircraft, the effective spring constant may depend greatly upon the firmness with which the pilot holds the control column.

7 Single-Degree-of-Freedom Flutter

Classical flutter, described in the previous chapter, assumed totally attached flow and coupling among at least two degrees of freedom. By comparison, single-degree-of-freedom flutter is driven primarily by separated-flow aerodynamics.

7.1 "GALLOPING" FLUTTER

This analysis follows that by J. P. Den Hartog, as described in *Mechanical Vibrations*, McGraw-Hill, New York, 1956:

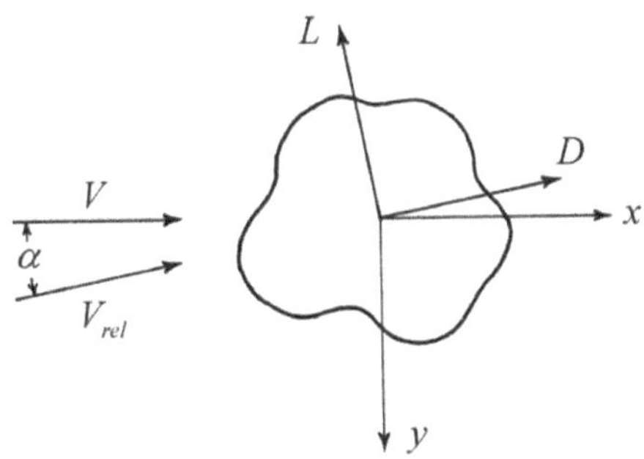

This is a form of flutter that occurs most commonly with long slender compliant structures, such as transmission lines, structural support cables, or tall skinny columns. Also, this type of flutter does not occur with perfectly-circular cross sections. Instead, it depends greatly upon the irregularity of the cross section, such as caused by being a prismatic shape, stranded cable, or from ice accretion. Consider an element of such a structure that is elastically supported. Because of the cross section's irregularity, both lift and drag are generated as functions of a:

$$L = \frac{\rho V_{rel}^2}{2} C_L(\alpha)\bar{d},\ D = \frac{\rho V_{rel}^2}{2} C_D(\alpha)\bar{d} \tag{7.1}$$

where $\bar{d}$ is the "defined diameter" of the cross section (an average diameter may be used, for example) and

$$V_{rel} = \left(V^2 + \dot{y}^2\right)^{1/2} = V \sec\alpha \tag{7.2}$$

$$\alpha = \tan^{-1}\left(\dot{y}/V\right) \tag{7.3}$$

DOI: 10.1201/9781032709093-7

The force in the cross-wind direction (y direction) is

$$F_y = -L\cos\alpha - D\sin\alpha$$

so one obtains from Equations (7.1) and (7.2) that

$$F_y = -\frac{\rho V^2}{2}\sec^2\alpha\left(C_L\cos\alpha + C_D\sin\alpha\right)\bar{d} = \frac{\rho V^2}{2}C_F\,\bar{d} \tag{7.4}$$

where

$$C_F = -\sec\alpha\left(C_L + C_D\tan\alpha\right) \tag{7.5}$$

Now, upon treating the elastically supported element as a simple spring-mass-dashpot system, one has the dynamic equation:

$$F_y = M\ddot{y} + C\dot{y} + Ky = M\left(\ddot{y} + 2\zeta\omega_n\dot{y} + \omega_n^2 y\right) \tag{7.6}$$

where, as described in Section 6.2.3, ω_n is the natural frequency and ζ is the damping ratio.

The criterion for galloping instability is that the system just sustains simple harmonic motion for small perturbations. Therefore, α may be assumed to be small, and Equations (7.4), (7.5), and (7.2) combine to give

$$F_Y \approx \left(\frac{\partial F_Y}{\partial\alpha}\right)_{\alpha=0}\times\alpha = \frac{\rho V^2}{2}\bar{d}\left(\frac{\partial C_F}{\partial\alpha}\right)_{\alpha=0}\times\alpha = -\frac{\rho V^2}{2}\bar{d}\left(\frac{\partial C_L}{\partial\alpha} + C_D\right)\frac{\dot{y}}{V}$$

This may be combined with Equation (7.6) to give

$$M\ddot{y} + \left[2M\zeta\omega_n + \frac{\rho V}{2}\bar{d}\left(\frac{\partial C_L}{\partial\alpha} + C_D\right)\right]\dot{y} + M\omega_n^2 y = 0$$

Clearly, whether the system is stable or not depends on the damping term:

$$2M\zeta\omega_n + \frac{\rho V}{2}\bar{d}\left(\frac{\partial C_L}{\partial\alpha} + C_D\right)$$

The damping ratio, ζ, is always positive, so the condition for neutral stability or instability is

$$\left(\frac{\partial C_L}{\partial\alpha} + C_D\right) \le \frac{4M\zeta\omega_n}{\rho V\bar{d}} \equiv A$$

In fact, since A is always positive and usually small, one may state that the condition for marginal or total instability is

$$\frac{\partial C_L}{\partial\alpha} + C_D \prec 0 \tag{7.7}$$

This is known as the "Den Hartog criterion," and it gives the interesting consequence that the lift-curve slope must be *negative* for instability. This is a strange notion for one used to the attached-flow aerodynamics of airfoils, where C_{L_α} is > 0. However, bear in mind that the cylinder has considerable separation in a pattern that is generally asymmetric with the x axis. This can give behavior such as shown by the C_L and C_D vs. α graph from R. H. Scanlan and R. L. Wardlaw ("Reduction of Flow-Induced Structural Vibrations," Section 2 of *Isolation of Mechanical Vibrations, Impact, and Noise*, AMD vol. 1, 1973, American Society of Mechanical Engineers, New York):

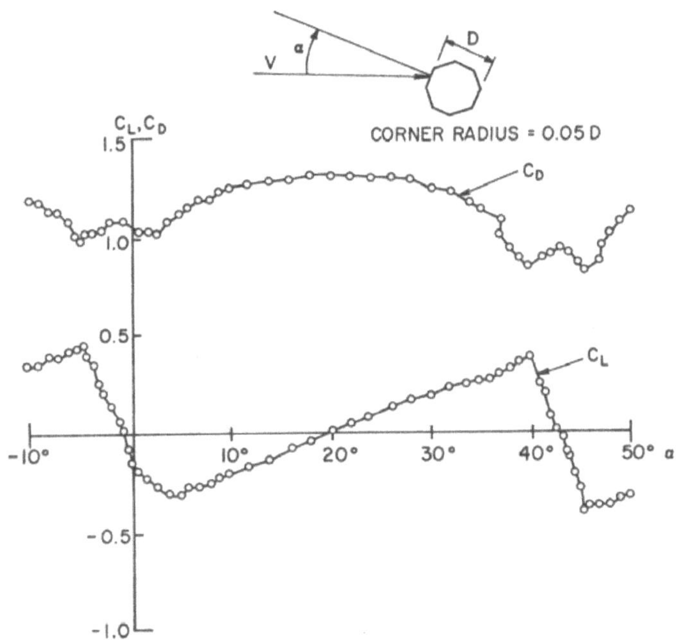

7.2 AEOLIAN FLUTTER

The galloping flutter, previously described, assumed a quasi-steady aerodynamic model. However, many cylinder shapes, including circular, experience alternating shed vortices at certain wind speeds.

V. Strouhal ("Uber eine besondere Art der Tonerregung," *Annalen der Physik, Leipzig*, vol. 5, 1878) determined a non-dimensional parameter that gives the relation between vortex shedding frequency, n_s (cycles per second), and wind speed:

$$S = \frac{n_s \bar{d}}{V}$$

where, in honor of this discovery, S is called the "Strouhal number." S can vary from 0.1 to 0.3 for various geometric shapes, with $S \approx 0.2$ for circular cylinders.

The alternating vortices cause an alternating asymmetrical transverse pressure pattern and, therefore, a transverse force alternating at frequency, n_s. If a structural mode has a frequency close to this, then Aeolian vibrations can lock in and persist even if the frequency subsequently varies somewhat from its original value. Measures to suppress this can include damping and breaking up of the regular vortex wake with surface features on the cylinder (spiral-wound strips, dimples, etc.).

As a historical note, the ancient Greeks designed harps that, when set into the wind, would produce different Aeolian tones from different strings, depending on the wind speed. These, naturally, are called Aeolian harps.

7.3 PLUNGING STALL FLUTTER

Flutter may occur when the airfoil is close to, or within, a stalled condition. This will be explained with an analysis developed from an idealized physical model presented in *A Modern Course in Aeroelasticity* by E. H. Dowell et al. (Sijthoff & Noordhoff, 1978).

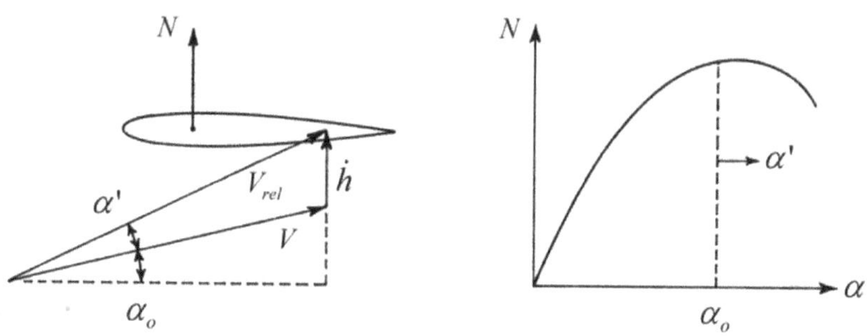

The normal force, N, is given by

$$N = \frac{\rho V_{rel}^2}{2} c C_N(\alpha_o, \alpha')$$

where

$$V_{rel}^2 = V^2 + 2V\dot{h}\sin\alpha_0 + \dot{h}^2$$

Also, $C_N(\alpha_o, \alpha')$ is assumed to be represented by a polynomial:

$$C_N = \sum_{0}^{K} C_i(\alpha_o)(\alpha')^n$$

where α_o is the reference (static) angle of attack and α' is the excursion from this:

$$\alpha' = \tan^{-1}\left(\tan\alpha_o + \frac{\dot{h}}{V\cos\alpha_o}\right) - \alpha_o$$

Choose simple harmonic motion for h:

$$h = h_o \cos\omega t \rightarrow \dot{h} = -\omega h_o \sin\omega t$$

The work done by the force, N, over one full cycle of motion determines whether the system will be stable or unstable:

$$Work\,/\,Cycle = \int_0^T N\dot{h}\,dt = \frac{1}{\omega}\int_0^{2\pi} N\dot{h}\,d(\omega t)$$

One may also obtain the power generated by the force and the airfoil's motion:

$$P = (Work\,/\,Cycle)(Cycle\,/\,Sec) = \frac{1}{2\pi}\int_0^{2\pi} N\dot{h}\,d(\omega t)$$

7.3.1 Numerical Example

Consider a section whose C_N behavior may be represented by

$$C_N = C_0 + C_1\alpha' + C_2(\alpha')^2$$

where

$$C_0 = -0.0144 + 0.1171\alpha_o - 0.0036(\alpha_o)^2$$

$$C_1 = 0.1183 - 0.0066\alpha_o - 0.00005(\alpha_o)^2$$

$$C_2 = -0.0037 - 0.00002\alpha_o - 0.000001(\alpha_o)^2$$

The other properties are

$$\rho = 1.225\,kg/m^3,\ c = 1.0\,m,\ h_o = 0.5\,m,\ \omega = 2.0\,rad/\sec,\ V = 50\,m/\sec$$

which give the power results for various values of static angle of attack, α_o:

α_o (Deg.)	P (Watts)
0	–103.81
2	–92.22
4	–80.59
6	–68.90
8	–57.09
10	–45.11
12	–32.94
14	–20.52
16	–7.83
18	5.16
20	18.49

From this it is seen that

$$(\alpha_o)_{flutter} = 17.21^\circ$$

This result is, in fact, not strongly dependent on the values for ω, V, and h_o. Instead, the primary reason for the instability is the negative slope of the $C_N(\alpha_o, \alpha')$ curve when α_o is above 16°. This is a situation where the α' excursions into the stalled region put energy into the system. In this respect, the significance of the negative slope is that it produces a flutter mechanism identical to that for galloping.

Note that the assumed aerodynamic model has been quasi-steady. It does not account for the significant effects of dynamic stall delay, hysteresis, etc. However, a small step toward a more realistic representation is to introduce a time-delay term, ψ, so that

$$h = h_o \cos(\omega t - \psi) \rightarrow \dot{h} = -\omega h_o \sin(\omega t - \psi)$$

The numerical example then gives

ψ (rad)	$(\alpha_o)_{flutter}$ (deg)
0	17.21
0.1	17.22
0.2	17.24
0.3	17.27
0.4	17.32
0.5	17.39

It is seen that there is not a great dependence on ψ, even for unrealistically large values. However, this is not the case for pitching stall flutter.

7.4 PITCHING STALL FLUTTER

Consider a section whose motion is constrained to pitching about a fixed point:

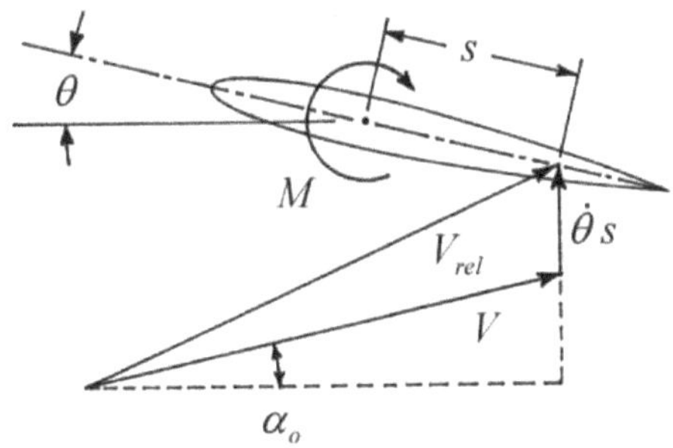

The aerodynamic moment about the pitching point is

$$M = \frac{\rho V_{rel}^2}{2} c^2 C_M\left(\alpha_o, \alpha'\right)$$

where

$$V_{rel}^2 = V^2 + 2V\dot{\theta}s \sin\alpha_o + \left(s\dot{\theta}\right)^2$$

$$\alpha' = \theta + \tan^{-1}\left(\tan\alpha_o + \frac{s\dot{\theta}}{V\cos\alpha_o}\right) - \alpha_o$$

Further, C_M will be considered to be due to the normal force, N, acting at the aerodynamic center:

$$C_M = C_N \times s_1$$

where s_1 is the distance between the aerodynamic center and the pitching point (positive when the *ac* is forward of the pitching point).

It is important to note that this C_M equation only exactly applies for totally attached flow. For example, separated regions modify not only the magnitude of C_N but also the location of its pressure center. Therefore, the following development should be considered to be a heuristic exercise.

The motion is given as

$$\theta = \theta_o \sin\left(\omega t - \psi\right) \rightarrow \dot{\theta} = \omega\theta_o \cos\left(\omega t - \psi\right)$$

and upon following a similar development to that for the plunging airfoil, the aerodynamic power generated by the section is

$$P = \frac{1}{2\pi}\int_0^{2\pi} M\dot{\theta}d\left(\omega t\right)$$

7.4.1 Numerical Example

Along with the previous values, choose

$$s_1 = 0.2\ m,\ s = 0.3\ m,\ \theta_o = 0.1\ rad,\ \omega = 2.0\ rad/\text{sec},\ V = 50\ m/\text{sec}$$

For $\psi = 0$, the following table shows P for various values of the static angle of attack, α_o:

α_o (Deg)	P (Watts)
0	0.249
4	0.198
8	0.152
12	0.108
16	0.061
20	0.005
24	−0.064

Unlike the plunging flutter case, the section is unstable for α_o below a certain value (20.3° in this case) and stable above this. This denotes a limit-cycle behavior. Also, $(\alpha_o)_{flutter}$ does not vary with V, nor does it change significantly with moderate variations in ω and θ_o. However, ψ has a strong effect, as shown as follows:

ψ (rad)	$(\alpha_o)_{flutter}$ (deg)*
0	20.3
0.002	21.3
0.004	22.7
0.006	24.8
0.008	27.8
0.010	31.6
0.012	36.0
0.014	40.6

* Values above which the system is stable.

Note how the phase angle is much more significant for the pitching case than for plunging stall flutter.

Appendix A
Lumped-Parameter Coupled Bending and Twisting

A.1 UNSWEPT WING

Consider a section of a finite wing where the mass center is displaced from the elastic center:

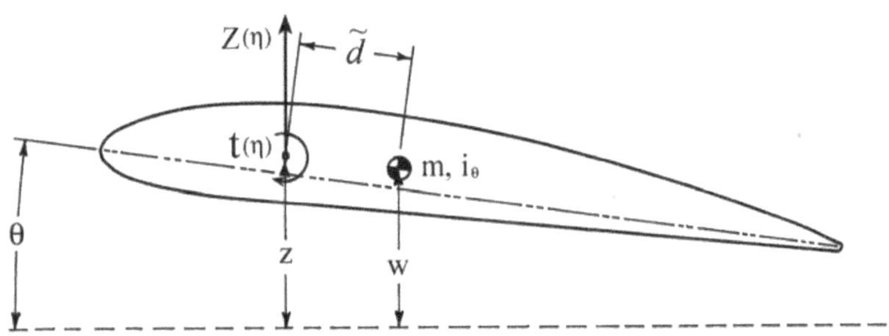

The displacement, $z(y)$, is given by

$$z(y) = \int_0^L C^{zz}(y,\eta) Z(\eta) d\eta \tag{A.1}$$

where the vertical force is given by

$$Z(\eta) = -m(\eta)\ddot{w}(\eta) = -m\left(\ddot{z} - \ddot{\theta}\tilde{d}\right)$$

Let

$$z(\eta) = \overline{z}(\eta) \sin \omega t,\ \theta(\eta) = \overline{\theta}(\eta) \sin \omega t \tag{A.2}$$

so Equation (A.1) becomes

$$\overline{z}(y) = \omega^2 \int_0^L C^{zz}(y,\eta) m(\eta)\left[\overline{z}(\eta) - \overline{\theta}(\eta)\tilde{d}\right] d\eta \tag{A.3}$$

With the following defined terms:

$$\overline{z}(y) \equiv \overline{z}_i \ , \overline{z}(\eta) \equiv \overline{z}_j \ , \overline{\theta}(\eta) \equiv \overline{\theta}_j \ , C^{zz}(y,\eta) \equiv C_{ij}^{zz} \ , m_j \equiv \int_{segment} m(\eta)d\eta \ , \tilde{d}(\eta) = \tilde{d}_j$$

Equation (A.3) may be written in the lumped element form:

$$\overline{z}_i = \omega^2 \sum_{j=1}^{n} C_{ij}^{zz} m_j \overline{z}_j - \omega^2 \sum_{j=1}^{n} C_{ij}^{zz} m_j \theta_j \tilde{d}_j$$

This may also be expressed in matrix form as

$$\{\overline{z}\} = \omega^2 \left[C^{zz}\right][m]\{\overline{z}\} - \omega^2 \left[C^{zz}\right][m]\left[\tilde{d}\right]\{\theta\} \tag{A.4}$$

where $[m]$ and $\left[\tilde{d}\right]$ are diagonal matrices.

Next, the twisting angle, θ, is given by

$$\theta(y) = \int_0^L C^{\theta\theta}(y,\eta)t(\eta)d\eta \tag{A.5}$$

where the twisting moment is given by

$$t(\eta) = -i_\theta(\eta)\ddot{\theta}(\eta) + m(\eta)\ddot{w}(\eta)\tilde{d}(\eta) = -i_\theta \ddot{\theta} + m\left(\ddot{z} - \ddot{\theta}\tilde{d}\right)\tilde{d}$$

With Equation (A.2), this becomes

$$t(\eta) = \omega^2 i_\theta \overline{\theta} \sin\omega t - \omega^2 m\left(\overline{z} - \overline{\theta}\tilde{d}\right)\tilde{d} \sin\omega t$$

so that, into Equation (A.5), one obtains

$$\overline{\theta}(y) = \omega^2 \int_0^L C^{\theta\theta}(y,\eta)\left\{i_\theta(\eta)\overline{\theta}(\eta) - m(\eta)\left[\overline{z}(\eta) - \overline{\theta}(\eta)\tilde{d}(\eta)\right]\tilde{d}(\eta)\right\}d\eta \tag{A.6}$$

As before, upon defining

$$\overline{\theta}(y) \equiv \overline{\theta}_i, \overline{\theta}(\eta) \equiv \overline{\theta}_j, C^{\theta\theta}(y,\eta) \equiv C_{ij}^{\theta\theta}, (i_\theta)_j \equiv \int_{segment} i_\theta(\eta)d\eta$$

the lumped parameter version of Equation (A.6) is

$$\overline{\theta}_i = \omega^2 \sum_{j=1}^{n} C_{ij}^{\theta\theta} (i_\theta)_j \overline{\theta}_j - \omega^2 \sum_{j=1}^{n} C_{ij}^{\theta\theta} m_j \tilde{d}_j \overline{z}_j + \omega^2 \sum_{j=1}^{n} C_{ij}^{\theta\theta} m_j \tilde{d}_j^2 \overline{\theta}_j$$

In matrix form, this becomes

$$\{\overline{\theta}\} = \omega^2 \left[C^{\theta\theta}\right]\left[i_\theta\right]\{\overline{\theta}\} - \omega^2 \left[C^{\theta\theta}\right][m]\left[\tilde{d}\right]\{\overline{z}\} + \omega^2 \left[C^{\theta\theta}\right][m]\left[\tilde{d}\right]\left[\tilde{d}\right]\{\overline{\theta}\}$$

which rearranges to give

$$\left\{\overline{\theta}\right\}=\omega^2\left[C^{\theta\theta}\right]\left[\left[i_\theta\right]+\left[m\right]\left[\tilde{d}\right]\left[\tilde{d}\right]\right]\left\{\overline{\theta}\right\}-\omega^2\left[C^{\theta\theta}\right]\left[m\right]\left[\tilde{d}\right]\left\{\overline{z}\right\} \tag{A.7}$$

As with $\left[m\right]$ and $\left[\tilde{d}\right]$, $\left[i_\theta\right]$ is also a diagonal matrix.

For Equations (A.4) and (A.7), define

$$\left[A\right]\equiv\left[C^{zz}\right]\left[m\right],\left[B\right]\equiv\left[C^{zz}\right]\left[m\right]\left[\tilde{d}\right]$$

$$\left[C\right]\equiv\left[C^{\theta\theta}\right]\left[\left[i_\theta\right]+\left[m\right]\left[\tilde{d}\right]\left[\tilde{d}\right]\right],\left[D\right]\equiv\left[C^{\theta\theta}\right]\left[m\right]\left[\tilde{d}\right]$$

so that one has

$$\left\{\overline{z}\right\}=\omega^2\left[A\right]\left\{\overline{z}\right\}-\omega^2\left[B\right]\left\{\overline{\theta}\right\}$$

$$\left\{\overline{\theta}\right\}=\omega^2\left[C\right]\left\{\overline{\theta}\right\}-\omega^2\left[D\right]\left\{\overline{z}\right\}$$

These can be combined to give

$$\begin{Bmatrix}\overline{z}\\ \overline{\theta}\end{Bmatrix}=\omega^2\begin{bmatrix}\left[A\right] & -\left[B\right]\\ -\left[D\right] & \left[C\right]\end{bmatrix}\begin{Bmatrix}\overline{z}\\ \overline{\theta}\end{Bmatrix}\equiv\omega^2\left[U\right]\begin{Bmatrix}\overline{z}\\ \overline{\theta}\end{Bmatrix} \tag{A.8}$$

which may be rewritten as

$$\left[\omega^2 U-I\right]\begin{Bmatrix}\overline{z}\\ \overline{\theta}\end{Bmatrix}=0$$

Upon defining $\lambda\equiv 1/\omega^2$, the eigenvalue problem is obtained:

$$\left[U-\lambda I\right]\begin{Bmatrix}\overline{z}\\ \overline{\theta}\end{Bmatrix}=0 \tag{A.9}$$

Numerical Example 1

Consider the Goland wing again, from Section 6.4.2. As before, one has the uniform spanwise properties of

$$\begin{aligned}EI\left(y_i\right)&=9.7523\times10^6\ N-m^2,\ GJ\left(y_i\right)\\ &=9.8763\times10^5\ N-m^2/rad,\ \tilde{d}\left(y_i\right)=0.1826\ m\end{aligned}$$

and the lumped spanwise properties of

$$m_i=21.7680\ kg,\left(i_{ea}\right)_i=5.2673\ kg-m^2,\left(i_\theta\right)_i=\left(i_{ea}\right)_i-m_i\,\tilde{d}_i^2$$

where the stations y_i are

$$y_i = 0.3048, 0.9144, 1.5240, 2.1336, 2.7432, 3.3528, 3.9624, 4.5720, 5.1816, 5.7912\ (m)$$

From Equation (A.9), one obtains the modal shapes and frequencies. The *first three* modal shapes are

$$\begin{Bmatrix} \bar{Z} \\ \bar{\theta} \end{Bmatrix}_{1,2,3} = \begin{bmatrix} -0.0026 & -0.0005 & -0.0041 \\ -0.0223 & -0.0048 & -0.0306 \\ -0.0589 & -0.0129 & -0.0674 \\ -0.1094 & -0.0246 & -0.0998 \\ -0.1711 & -0.0393 & -0.1172 \\ -0.2412 & -0.0567 & -0.1138 \\ -0.3172 & -0.0759 & -0.0889 \\ -0.3971 & -0.0965 & -0.0463 \\ -0.4790 & -0.1179 & 0.0078 \\ -0.5618 & -0.1396 & 0.0669 \\ 0.0099 & -0.0349 & 0.0913 \\ 0.0299 & -0.1038 & 0.2558 \\ 0.0494 & -0.1697 & 0.3671 \\ 0.0683 & -0.2308 & 0.3984 \\ 0.0860 & -0.2856 & 0.3393 \\ 0.1020 & -0.3328 & 0.1984 \\ 0.1160 & -0.3714 & 0.0032 \\ 0.1272 & -0.4006 & -0.2054 \\ 0.1352 & -0.4201 & -0.3815 \\ 0.1395 & -0.4298 & -0.4835 \end{bmatrix}$$

and the corresponding frequencies are

$$\omega_1 = 48.215,\ \omega_2 = 95.631,\ \omega_3 = 242.43\ (rad/\text{sec})$$

Numerical Example 2

It is also of interest to consider a coupled structure with variable spanwise properties, such as described in Example 5.1. For this case, the spanwise parameters are

$$EI(y_i) = 9.7523 \times 10^6 \left[1 - 0.9(y_i/L)^{0.25}\right] \left(N - m^2\right)$$

$$GJ(y_i) = 9.8763 \times 10^5 \left[1 - 0.9(y_i/L)^{0.25}\right] \left(N - m^2/rad\right)$$

$$\tilde{d}(y_i) = 0.1826\left[1 - 0.5(y_i / L)\right](m)$$

$$m_i = \left(35.7087\left[1 - 0.8(y_i / L)^{0.5}\right]\right) \times \delta y_i \,(kg)$$

$$(i_{ea})_i = 8.6406\left[1 - 0.8(y_i / L)^{0.33}\right] \times \delta y_i \,(kg - m^2)$$

$$(i_\theta)_i = (i_{ea})_i - m_i \tilde{d}_i^2$$

where the segment widths, $\delta y_i = 0.6096\ m$, and the y_i stations are the same as before. The solution of Equation (A.9) gives the following first three modes:

$$\begin{Bmatrix} \bar{Z} \\ \bar{\theta} \end{Bmatrix}_{1,2,3} = \begin{bmatrix} -0.0019 & 0.0003 & -0.0020 \\ -0.0166 & 0.0026 & -0.0158 \\ -0.0468 & 0.0077 & -0.0372 \\ -0.0923 & 0.0157 & -0.0582 \\ -0.1520 & 0.0268 & -0.0706 \\ -0.2239 & 0.0409 & -0.0680 \\ -0.3056 & 0.0574 & -0.0478 \\ -0.3944 & 0.0760 & -0.0110 \\ -0.4874 & 0.0959 & 0.0378 \\ -0.5823 & 0.1163 & 0.0923 \\ 0.0059 & 0.0205 & 0.0569 \\ 0.0194 & 0.0669 & 0.1763 \\ 0.0359 & 0.1220 & 0.2863 \\ 0.0542 & 0.1827 & 0.3500 \\ 0.0738 & 0.2459 & 0.3376 \\ 0.0935 & 0.3087 & 0.2343 \\ 0.1123 & 0.3679 & 0.0047 \\ 0.1287 & 0.4198 & -0.1931 \\ 0.1415 & 0.4599 & -0.4283 \\ 0.1489 & 0.4833 & -0.5852 \end{bmatrix}$$

And the corresponding frequencies are

$$\omega_1 = 58.068,\ \omega_2 = 97.238,\ \omega_3 = 210.83\,(rad/\mathrm{sec})$$

Upon comparison with the first three frequencies of the assumed-mode solution of Example 5.3, it is seen that the values correspond very well (as would be expected).

A.2 SWEPT WING

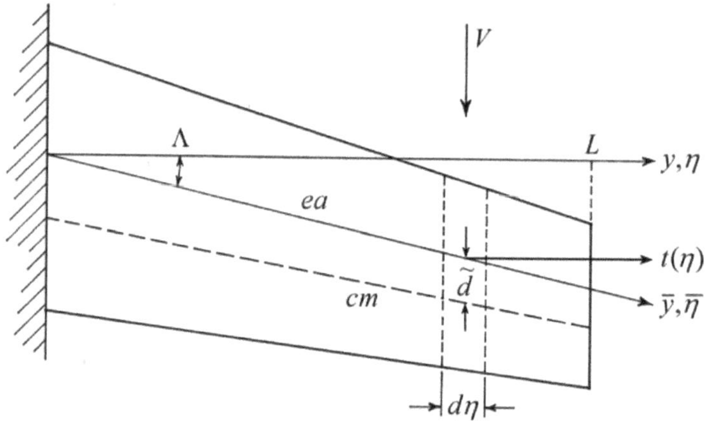

For this case, the displacement and rotation equations include the coupled terms, $C^{z\theta}$ and $C^{\theta z}$:

$$z(y)=\int_0^L C^{zz}(y,\eta)Z(\eta)d\eta+\int_0^L C^{z\theta}(y,\eta)t(\eta)d\eta \tag{A.10}$$

$$\theta(y)=\int_0^L C^{\theta\theta}(y,\eta)t(\eta)d\eta+\int_0^L C^{\theta z}(y,\eta)Z(\eta)d\eta \tag{A.11}$$

where, from before,

$$Z(\eta)=-m\left(\ddot{z}-\ddot{\theta}\tilde{d}\right),\ t(\eta)=-i_\theta\ddot{\theta}+m\left(\ddot{z}-\ddot{\theta}\tilde{d}\right)\tilde{d}$$

Also,

$$z(\eta)=\bar{z}(\eta)\sin\omega t,\ \theta(\eta)=\bar{\theta}(\eta)\sin\omega t$$

so that Equations (A.10) and (A.11) become

$$\begin{aligned}\bar{z}(y)&=\omega^2\int_0^L C^{zz}(y,\eta)m(\eta)\left[\bar{z}(\eta)-\bar{\theta}(\eta)\tilde{d}\right]d\eta\\&+\omega^2\int_0^L C^{z\theta}(y,\eta)\left\{i_\theta\bar{\theta}(\eta)-m(\eta)\left[\bar{z}(\eta)-\bar{\theta}(\eta)\tilde{d}\right]\tilde{d}\right\}d\eta\end{aligned} \tag{A.12}$$

$$\bar{\theta}(y)=\omega^2\int_0^L C^{\theta\theta}(y,\eta)\left\{i_\theta\bar{\theta}(\eta)-m(\eta)\left[\bar{z}(\eta)-\bar{\theta}(\eta)\tilde{d}\right]\tilde{d}\right\}d\eta$$
$$+\omega^2\int_0^L C^{\theta z}(y,\eta)m(\eta)\left[\bar{z}(\eta)-\bar{\theta}(\eta)\tilde{d}\right]d\eta \tag{A.13}$$

As before, define lumped parameter terms

$$\bar{z}(y)\equiv\bar{z}_i,\ \bar{z}(\eta)\equiv\bar{z}_j,\ \bar{\theta}(y)\equiv\bar{\theta}_i,\ \bar{\theta}(\eta)\equiv\bar{\theta}_j,\ C^{zz}(y,\eta)\equiv C_{ij}^{zz},\ C^{z\theta}(y,\eta)\equiv C_{ij}^{z\theta}$$

$$C^{\theta z}(y,\eta)\equiv C_{ij}^{\theta z},\ C^{\theta\theta}(y,\eta)\equiv C_{ij}^{\theta\theta},\ m_j\equiv\int_{segment} m(\eta)d\eta,\ (i_\theta)_j\equiv\int_{segment} i_\theta(\eta)d\eta$$

$$\tilde{d}(\eta)=\tilde{d}_j$$

so that Equations (A.12) and (A.13) become

$$\bar{z}_i=\omega^2\sum_{j=1}^n C_{ij}^{zz}m_j\bar{z}_j-\omega^2\sum_{j=1}^n C_{ij}^{zz}m_j\bar{\theta}_j\tilde{d}_j+\omega^2\sum_{j=1}^n C_{ij}^{z\theta}(i_\theta)_j\bar{\theta}_j$$
$$-\omega^2\sum_{j=1}^n C_{ij}^{z\theta}m_j\bar{z}_j\tilde{d}_j+\omega^2\sum_{j=1}^n C_{ij}^{z\theta}m_j\bar{\theta}_j\tilde{d}_j^2 \tag{A.14}$$

$$\bar{\theta}_i=\omega^2\sum_{j=1}^n C_{ij}^{\theta\theta}(i_\theta)_j\bar{\theta}_j-\omega^2\sum_{j=1}^n C_{ij}^{\theta\theta}m_j\bar{z}_j\tilde{d}_j+\omega^2\sum_{j=1}^n C_{ij}^{\theta\theta}m_j\tilde{d}_j^2\bar{\theta}_j$$
$$+\omega^2\sum_{j=1}^n C_{ij}^{\theta z}m_j\bar{z}_j-\omega^2\sum_{j=1}^n C_{ij}^{\theta z}m_j\bar{\theta}_j\tilde{d}_j \tag{A.15}$$

In matrix form, Equations (A.14) and (A.15) are

$$\{\bar{z}\}=\omega^2\left[C^{zz}\right][m]\{\bar{z}\}-\omega^2\left[C^{zz}\right][m]\left[\tilde{d}\right]\{\bar{\theta}\}+\omega^2\left[C^{z\theta}\right][i_\theta]\{\bar{\theta}\}$$
$$-\omega^2\left[C^{z\theta}\right][m]\left[\tilde{d}\right]\{\bar{z}\}+\omega^2\left[C^{z\theta}\right][m]\left[\tilde{d}\right]\left[\tilde{d}\right]\{\bar{\theta}\} \tag{A.16}$$

$$\{\bar{\theta}\}=\omega^2\left[C^{\theta\theta}\right][i_\theta]\{\bar{\theta}\}-\omega^2\left[C^{\theta\theta}\right][m]\left[\tilde{d}\right]\{\bar{z}\}+\omega^2\left[C^{\theta\theta}\right][m]\left[\tilde{d}\right]\left[\tilde{d}\right]\{\bar{\theta}\}$$
$$+\omega^2\left[C^{\theta z}\right][m]\{\bar{z}\}-\omega^2\left[C^{\theta z}\right][m]\left[\tilde{d}\right]\{\bar{\theta}\} \tag{A.17}$$

Now, define

$$\left[\tilde{A}\right]\equiv\left[C^{zz}\right][m]-\left[C^{z\theta}\right][m]\left[\tilde{d}\right]$$

$$\left[\tilde{B}\right] \equiv -\left[C^{zz}\right]\left[m\right]\left[\tilde{d}\right] + \left[C^{z\theta}\right]\left[m\right]\left[\tilde{d}\right]\left[\tilde{d}\right] + \left[C^{z\theta}\right]\left[i_\theta\right]$$

$$\left[\tilde{C}\right] \equiv \left[C^{\theta\theta}\right]\left[i_\theta\right] + \left[C^{\theta\theta}\right]\left[m\right]\left[\tilde{d}\right]\left[\tilde{d}\right] - \left[C^{\theta z}\right]\left[m\right]\left[\tilde{d}\right]$$

$$\left[\tilde{D}\right] \equiv -\left[C^{\theta\theta}\right]\left[m\right]\left[\tilde{d}\right] + \left[C^{\theta z}\right]\left[m\right]$$

so that Equations (A.16) and (A.17) become

$$\left\{\overline{z}\right\} = \omega^2\left[\tilde{A}\right]\left\{\overline{z}\right\} + \omega^2\left[\tilde{B}\right]\left\{\overline{\theta}\right\}$$

$$\left\{\overline{\theta}\right\} = \omega^2\left[\tilde{C}\right]\left\{\overline{\theta}\right\} + \omega^2\left[\tilde{D}\right]\left\{\overline{z}\right\}$$

which combine to give

$$\begin{Bmatrix}\overline{z}\\ \overline{\theta}\end{Bmatrix} = \omega^2\begin{bmatrix}\left[\tilde{A}\right] & \left[\tilde{B}\right]\\ \left[\tilde{D}\right] & \left[\tilde{C}\right]\end{bmatrix}\begin{Bmatrix}\overline{z}\\ \overline{\theta}\end{Bmatrix} \equiv \omega^2\left[\tilde{U}\right]\begin{Bmatrix}\overline{z}\\ \overline{\theta}\end{Bmatrix} \tag{A.18}$$

This may be rewritten as

$$\left[\omega^2\tilde{U} - I\right]\begin{Bmatrix}\overline{z}\\ \overline{\theta}\end{Bmatrix} = 0$$

So that, upon defining $\tilde{\lambda} \equiv 1/\omega^2$, one obtains the eigenvalue problem:

$$\left[\tilde{U} - \tilde{\lambda}I\right]\begin{Bmatrix}\overline{z}\\ \overline{\theta}\end{Bmatrix} = 0 \tag{A.19}$$

Numerical Example 3

Consider a swept version of the wing from Example 1. In this case, one has

$$EI\left(\overline{y}_i\right) = 9.7523\times10^6\,N - m^2,\; GJ\left(\overline{y}_i\right) = 9.8763\times10^5\,N - m^2/rad$$

$$\tilde{d}\left(y_i\right) = 0.1826/\cos\Lambda\left(m\right),\; m_i = 21.7680\,kg,\; \left(i_{ea}\right)_i = 5.2673/\cos^2\Lambda\left(kg - m^2\right)$$

As before,

$$\left(i_\theta\right)_i = \left(i_{ea}\right)_i - m_i\tilde{d}_i^2$$

Further, the $\overline{y}_i$ stations along the elastic axis are

$$\overline{y}_i = 0.3048, 0.9144, 1.5240, 2.1336, 2.7432, 3.3528, 3.9624, 4.5720, 5.1816, 5.7912\left(m\right)$$

From Equation (A.19), one obtains the modal shapes and frequencies. If $\Lambda = 20^\circ$, then the *first three* modal shapes are

$$\begin{Bmatrix} \overline{Z} \\ \overline{\theta} \end{Bmatrix} = \begin{Bmatrix} -0.0025 & -0.0007 & -0.0046 \\ -0.0212 & -0.0062 & -0.0345 \\ -0.0561 & -0.0165 & -0.0770 \\ -0.1044 & -0.0309 & -0.1156 \\ -0.1633 & -0.0486 & -0.1376 \\ -0.2305 & -0.0690 & -0.1350 \\ -0.3036 & -0.0911 & -0.1058 \\ -0.3805 & -0.1144 & -0.0531 \\ -0.4594 & -0.1382 & 0.0161 \\ -0.5393 & -0.1620 & 0.0937 \\ 0.0154 & -0.0342 & 0.0964 \\ 0.0448 & -0.1018 & 0.2640 \\ 0.0723 & -0.1667 & 0.3714 \\ 0.0975 & -0.2271 & 0.3959 \\ 0.1200 & -0.2815 & 0.3310 \\ 0.1395 & -0.3285 & 0.1884 \\ 0.1555 & -0.3671 & -0.0035 \\ 0.1678 & -0.3965 & -0.2049 \\ 0.1762 & -0.4163 & -0.3726 \\ 0.1805 & -0.4261 & -0.4688 \end{Bmatrix}$$

and the corresponding frequencies are

$$\omega_1 = 46.834,\ \omega_2 = 98.571,\ \omega_3 = 231.84\ (rad/\text{sec})$$

These are close to those for Example 1. Note that sweeping a beam would affect the mode shapes because of the coordinate change, but it should not affect the modal frequencies.

Appendix B
Flutter of the Representative Section with Control Surface

The unsteady lift equation for a simply-hinged section, such as shown as follows, is given by Theodorsen (NACA TR. 496, 1935):

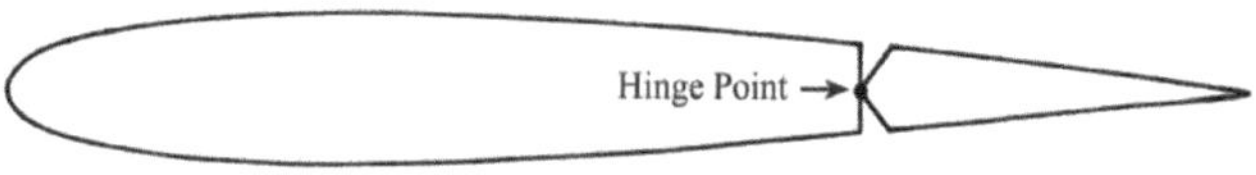

$$L' = L'_{circ} + L'_{app}$$

where the circulation term is

$$L'_{circ} = 2\pi\rho Vb\left[F(k) + iG(k)\right]\left[V\alpha + \dot{h} + b\left(\frac{1}{2} - a\right)\dot{\alpha} + \left(\frac{T_{10}V}{\pi}\right)\beta + \left(\frac{bT_{11}}{2\pi}\right)\dot{\beta}\right]$$

and the apparent mass term is

$$L'_{app} = \rho b^2\left(V\pi\dot{\alpha} + \pi\ddot{h} - \pi ba\ddot{\alpha} - VT_4\dot{\beta} - T_1 b\ddot{\beta}\right)$$

The T terms pertain to the geometry of the control surface:

$$T_1 = -1/3\left(1-c^2\right)^{1/2}\left(2+c^2\right) + c\cos^{-1}c$$

$$T_4 = -\cos^{-1}c + c\left(1-c^2\right)^{1/2}$$

$$T_{10} = \left(1-c^2\right)^{1/2} + \cos^{-1}c$$

$$T_{11} = (1-2c)\cos^{-1}c + \left(1-c^2\right)^{1/2}(2-c)$$

Recall that bc is the distance from the section's midchord to the hinge point. Also, this is related to the ratio between the control chord and the total section chord by

$$c = 1 - 2\frac{control\ chord}{total\ chord}$$

Further, the section moment equation is given by

$$M' = M'_{circ} + M'_{app}$$

where

$$M'_{circ} = -bxL'_{circ}$$

$$M'_{app} = \rho b^2 \left\{ \pi(1/2 - a)Vb\dot{\alpha} + \pi b^2 \left(1/8 + a^2\right)\ddot{\alpha} + (T_4 + T_{10})V^2\beta \right\}$$
$$+ \rho b^2 \left\{ [T_1 - T_8 - (c-a)T_4 + T_{11}/2]Vb\dot{\beta} - [T_7 + (c-a)T_1]b^2\ddot{\beta} - a\pi b\ddot{h} \right\}$$

and

$$T_7 = -\left(1/8 + c^2\right)\cos^{-1}c + c/8\left(1-c^2\right)^{1/2}\left(7+2c^2\right)$$

$$T_8 = -1/3\left(1-c^2\right)^{1/2}\left(2c^2+1\right) + c\cos^{-1}c$$

The hinge moment equation is

$$P' = P'_{circ} + P'_{app}$$

where

$$P'_{circ} = \frac{b}{2\pi}T_{12}L'_{circ}$$

$$P'_{app} = \rho b^2 \left\{ [-2T_9 - T_1 + T_4(a - 1/2)]\, Vb\dot{\alpha} + 2T_{13}b^2\ddot{\alpha} \right\}$$
$$+ \rho b^2 \left[\frac{V^2}{\pi}(T_5 - T_4T_{10})\beta - \frac{Vb}{2\pi}T_4T_{11}\dot{\beta} - \frac{T_3b^2}{\pi}\ddot{\beta} - T_1b\ddot{h} \right]$$

and

$$T_3 = -\left(1/8 + c^2\right)\left(\cos^{-1}c\right)^2 + c/4\left(1-c^2\right)^{1/2}\left(\cos^{-1}c\right)\left(7+2c^2\right) - 1/8\left(1-c^2\right)\left(5c^2+4\right)$$

$$T_5 = -\left(1-c^2\right) - \left(\cos^{-1}c\right)^2 + 2c\left(1-c^2\right)^{1/2}\cos^{-1}c$$

$$T_9 = 1/2\left\{1/3\left[\left(1-c^2\right)^{1/2}\right]^3 + aT_4\right\}$$

$$T_{12} = \left(1-c^2\right)^{1/2}(2+c) - (2c+1)\cos^{-1}c$$

$$T_{13} = 1/2\left[-T_7 - (c-a)T_1\right]$$

Now, upon writing the motion variables as

$$\alpha = \bar{\alpha}e^{i\omega t},\ h = \bar{h}e^{i\omega t},\ \beta = \bar{\beta}e^{i\omega t}$$

one may follow the method described previously for the fixed control section to now obtain

$$L' = \rho V^2 A'_{hh} h + \rho V^2 A'_{h\alpha}\alpha + \rho V^2 A'_{h\beta}\beta + \rho V A''_{hh}\dot{h} + \rho V A''_{h\alpha}\dot{\alpha} + \rho V A''_{h\beta}\dot{\beta}$$

$$M' = \rho V^2 A'_{\alpha h} h + \rho V^2 A'_{\alpha\alpha}\alpha + \rho V^2 A'_{\alpha\beta}\beta + \rho V A''_{\alpha h}\dot{h} + \rho V A''_{\alpha\alpha}\dot{\alpha} + \rho V A''_{\alpha\beta}\dot{\beta}$$

$$P' = \rho V^2 A'_{\beta h} h + \rho V^2 A'_{\beta\alpha}\alpha + \rho V^2 A'_{\beta\beta}\beta + \rho V A''_{\beta h}\dot{h} + \rho V A''_{\beta\alpha}\dot{\alpha} + \rho V A''_{\beta\beta}\dot{\beta}$$

where A'_{hh}, $A'_{h\alpha}$, A''_{hh}, $A''_{h\alpha}$, $A'_{\alpha h}$, $A'_{\alpha\alpha}$, $A''_{\alpha h}$, $A''_{\alpha\alpha}$ are the same as previously derived for Equations (6.18) and (6.19), and the new terms are

$$A'_{h\beta} = 2b\left[F(k)T_{10} - G(k)T_{11}k/2\right] + T_1 bk^2$$

$$A''_{h\beta} = 2b^2\left[F(k)T_{11}/2 + G(k)T_{10}/k\right] - b^2 T_4$$

$$A'_{\alpha\beta} = -bx\left\{2b\left[F(k)T_{10} - G(k)T_{11}k/2\right]\right\} + b^2\left(T_4 + T_{10}\right) + b^2\left[T_7 + (c-a)T_1\right]k^2$$

$$A''_{\alpha\beta} = -bx\left\{2b^2\left[F(k)T_{11}/2 + G(k)T_{10}/k\right]\right\} + b^3\left[T_1 - T_8 - (c-a)T_4 + T_{11}/2\right]$$

$$A'_{\beta h} = -bG(k)kT_{12} + bT_1 k^2$$

$$A'_{\beta\alpha} = b^2 T_{12}\left[F(k) - G(k)(1/2 - a)k\right] - 2b^2 T_{13}k^2$$

$$A'_{\beta\beta} = b^2 T_{12}\left[F(k)\frac{T_{10}}{\pi} - G(k)\frac{T_{11}}{2\pi}k\right] + \frac{b^2}{\pi}\left(T_5 - T_4 T_{10}\right) + \frac{b^2}{\pi}T_3 k^2$$

$$A''_{\beta h} = b^2 T_{12} F(k)$$

$$A''_{\beta\alpha} = b^3 T_{12}\left[F(k)(1/2 - a) + G(k)/k\right] + b^3\left[T_4(a - 1/2) - 2T_9 - T_1\right]$$

$$A''_{\beta\beta} = b^3 T_{12}\left[F(k)\frac{T_{11}}{2\pi} + G(k)\frac{T_{10}}{\pi k}\right] - \frac{b^3}{2\pi}T_4 T_{11}$$

The aerodynamic force and moments may be expressed as an expanded form of Equations (6.20) and (6.21):

$$\begin{Bmatrix} L' \\ M' \\ P' \end{Bmatrix} = \rho V^2 \begin{bmatrix} A'_{hh} & A'_{h\alpha} & A'_{h\beta} \\ A'_{\alpha h} & A'_{\alpha\alpha} & A'_{\alpha\beta} \\ A'_{\beta h} & A'_{\beta\alpha} & A'_{\beta\beta} \end{bmatrix} \begin{Bmatrix} h \\ \alpha \\ \beta \end{Bmatrix} + \rho V \begin{bmatrix} A''_{hh} & A''_{h\alpha} & A''_{h\beta} \\ A''_{\alpha h} & A''_{\alpha\alpha} & A''_{\alpha\beta} \\ A''_{\beta h} & A''_{\beta\alpha} & A''_{\beta\beta} \end{bmatrix} \begin{Bmatrix} \dot{h} \\ \dot{\alpha} \\ \dot{\beta} \end{Bmatrix}$$

$$\equiv \rho V^2 \left[A_S\right] \begin{Bmatrix} h \\ \alpha \\ \beta \end{Bmatrix} + \rho V \left[A_D\right] \begin{Bmatrix} \dot{h} \\ \dot{\alpha} \\ \dot{\beta} \end{Bmatrix}$$

where, as before, $\lfloor A_S \rfloor$ may be interpreted as an aerodynamic "spring" matrix and $\lfloor A_D \rfloor$ may be interpreted as an aerodynamic "damping" matrix. Upon incorporation into the complete dynamic equations, one obtains

$$\begin{bmatrix} M & S_\alpha & S_\beta \\ S_\alpha & I_\alpha & P_{\alpha\beta} \\ S_\beta & P_{\alpha_\beta} & I_\beta \end{bmatrix} \begin{Bmatrix} \ddot{h} \\ \ddot{\alpha} \\ \ddot{\beta} \end{Bmatrix} + \begin{bmatrix} M\omega_h^2 & & \\ & I_\alpha \omega_\alpha^2 & \\ & & I_\beta \omega_\beta^2 \end{bmatrix} \begin{Bmatrix} h \\ \alpha \\ \beta \end{Bmatrix} = -\rho V^2 \lfloor A_S \rfloor \begin{Bmatrix} h \\ \alpha \\ \beta \end{Bmatrix} - \rho V \lfloor A_D \rfloor \begin{Bmatrix} \dot{h} \\ \dot{\alpha} \\ \dot{\beta} \end{Bmatrix}$$

With the substitution of

$$\alpha = \bar{\alpha} e^{\lambda t}, \quad h = \bar{h} e^{\lambda t}, \quad \beta = \bar{\beta} e^{\lambda t}$$

the solution takes the form of an eigenvalue problem for the stability roots, $\lambda = \lambda_r + i\omega$, and the coresponding modal vectors. As before, only the flutter case ($\lambda_r = 0$) exactly satisfies the assumption of simple harmonic motion for the aerodynamic terms.

NUMERICAL EXAMPLE

Consider the fixed-control example in Section 6.3.1. This is now given a control surface whose chord is 30% of the section's total chord. Therefore,

$$c = 1 - 2(0.3) = 0.4$$

Further, for a meaningful comparison with the fixed-control example, the parameters will be chosen so that total mass, mass center, and total moment of inertia match the previous case:

$$m_w + m_\beta = M_{\mathit{fixed\ control}} \equiv M_{fc}$$

$$x_\alpha = \left[M_{fc} (x_\alpha)_{fc} - m_\beta (c - a + x_\beta) \right] / m_w$$

$$i_w + i_\beta = b^2 \left[M_{fc} (r_\alpha)^2_{fc} - m_w x_\alpha^2 - m_\beta (c - a + x_\beta)^2 \right]$$

As given before, the parameters are

$$b = 1.0, \; a = -0.1, \; x = a + 0.5 = 0.4, \; \omega_\alpha = 1.0, \; \omega_h = 0.5, \; (x_\alpha)_{fc} = 0.2$$

Also, the mass properties are

$$M_{fc} = 10\pi\rho b^2, \; (r_\alpha)^2_{fc} = 0.25$$

CASE 1: VARIATION OF CONTROL-SPRING STIFFNESS

Choose

$$m_w = 95\%M_{fc},\ m_\beta = 5\%M_{fc},\ x_\beta = 0.6$$

These give that $x_\alpha = 0.15263$ and $i_w + i_\beta = 0.16737M_{fc}$. Assume that the inertial terms vary in the same proportion as thc mass terms so that

$$i_w = 0.95\left(i_w + i_\beta\right) = 0.159M_{fc}, i_\beta = 0.05\left(i_w + i_\beta\right) = 0.00837M_{fc}$$

The parameter for determining the control spring stiffness is ω_β, which is varied from 1.0 to 10. The results for flutter speed parameter, $V_f/b\omega_\alpha$; flutter frequency ratio, ω_f/ω_α; and reduced flutter frequency, k_f, are shown here:

ω_β	**10.0**	**5.0**	**2.0**	**1.0**
$V_f/b\omega_\alpha$	1.294	1.292	1.276	0
ω_f/ω_α	0.800	0.798	0.778	...
k_f	0.619	0.618	0.610	...

This shows that flutter occurs suddenly when the control spring stiffness goes below a certain value.

CASE 2: VARIATION OF CONTROL-SURFACE MASS

For this case, choose the values of $\omega_\beta = 2.0$ and $x_\beta = 0.6$, and let m_β vary from 5% to 7% of M_{fc}. The results are

m_β	**5% M_{fc}**	**5.5% M_{fc}**	**6% M_{fc}**	**7% M_{fc}**
$V_f/b\omega_\alpha$	1.276	1.272	0	0
ω_f/ω_α	0.779	0.778	...	...
k_f	0.611	0.612	...	...

This clearly shows how a slight increase in the control-surface mass can cause an abrupt onset of flutter. In fact, for cable-actuated controls, it is standard practice to make these as light as possible. An example of this is shown in the following figure, which illustrates the typical construction of advanced aircraft for the 1930s and 1940s where, even for a metal-skinned structure, the control surfaces were open-framed and fabric-covered.

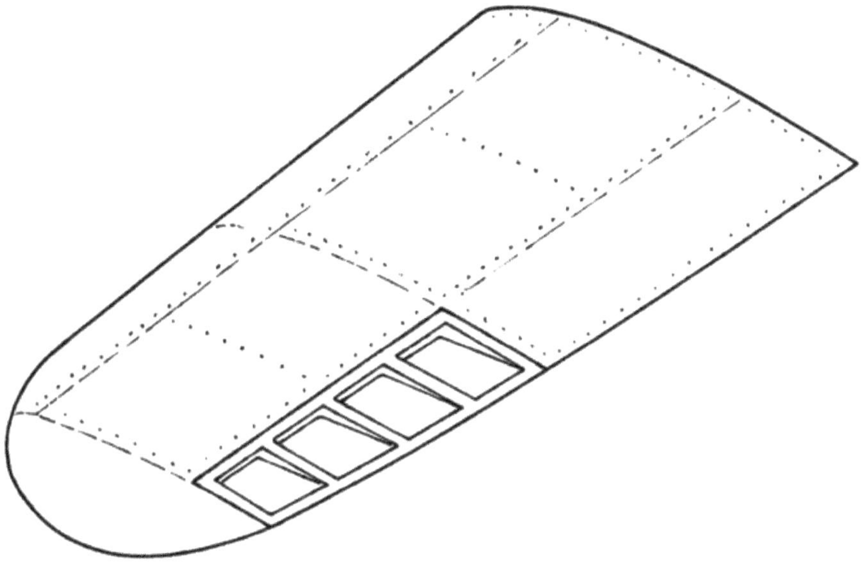

Case 3: Location of Control-Surface's Mass

For this case, choose $\omega_\beta = 2.0$, $m_w = 90\% M_{fc}$, $m_\beta = 10\% M_{fc}$. Likewise, i_w and i_β are apportioned the same. The control-surface mass-center location, x_β, is varied from 0 to 0.6:

x_β	**0**	**0.2**	**0.35**	**0.4**
$V_f/b\omega_\alpha$	1.314	1.289	1.270	0
ω_f/ω_α	0.773	0.777	0.779	...
k_f	0.589	0.603	0.613	...

This example shows the deleterious effect of locating the control-surface mass center too far aft. However, this also shows that flutter can be readily suppressed with a counterbalance, which has been a design feature for several aircraft. The drawing that follows shows one such example applied to the Lockheed P-38 "Lightning" fighter of the 1940s.

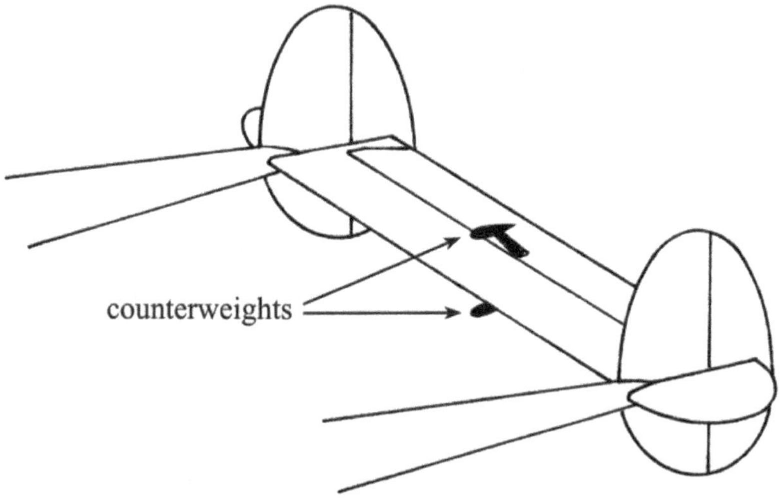

Index

A

aerodynamic terms
- AGARD representation, 119
- definition, 117
- finite wing, 142, 152
- representative section, 121
- section with control surface, 174

aeroelastic tailoring, 46
AeroVironment Span-loader airplane, 106
aileron, 48, 140, 155
anti-symmetrical modes, 78
apparent (virtual) mass, 118
assumed-mode method
- beam vibration, 83
- coupled bending and twisting vibration, 87
- divergence, 36
- finite wing, coupled, 148
- finite wing, uncoupled, 140
- lumped-parameter beam vibration, 106
- torsional vibration, 86

B

beam end conditions, 72
bending vibration, 71
- assumed-mode solution, 83
- cantilever beam, 73
- free-free beam, 75
- kinetic energy, 83

Boeing B-47, 41, 139

C

cantilever beam, 5
Castigliano's first theorem, 13
center-of-pressure, 50
- subsonic, 52
- supersonic, 53

Collar's triangle, 1
complementary energy, 13
control deflection, 48
control effectiveness, 50
control reversal
- definition, 49
- finite wing
  - dynamic, 57
  - static, 53
- representative section, 49

control surface, 48, 60, 174
counterbalance, 180
coupled bending and twisting
- swept wing, 170
- unswept wing, 87, 165

D

discrete segments /elements, 19
divergence
- representative section
  - dynamic, 129
  - static, 28
- swept wing
  - approximate equation, 46
  - discrete-element model, 44
  - semi-rigid model, 38
- unswept wing
  - approximate equation, 29
  - assumed-mode method, 36
  - integral method, 33
  - uniform, 29

E

eigenvalue equation
- assumed-mode vibration
  - bending, 84
  - coupled bending and twisting, 90
  - torsion, 85
- divergence, 35, 37, 45
- lumped-parameter model
  - cantilever bending, 93
  - cantilever torsion, 94
  - coupled bending and twisting, 167, 172
  - free-free bending, 102

elastic axis, 3
elastic center, 2
elastic equation
- bending, 15, 27
- torsion, 27

equivalent-spring parameters (semi-rigid wing), 43

F

flexibility influence coefficient, 5
flexibility influence-coefficient matrix
- definition, 6
- example, 19
- swept beam, 21

flutter, classical
- definition, 112
- energy interpretation, 139
- examples from literature, 136
- finite wing, 140
- iterative solution, 134
- Pines' model, 126
- representative section, 113, 121
- root coalescence, 124

section with control surface, 174
V-g Plots, 112, 135
flutter, non-classical
aeolian, 159
definition, 112, 157
galloping, 157
limit cycle, 164
stall
pitching, 163
plunging, 160
Fokker D-8, 33
forced-beam formulation, 107

G

generalized coordinate, 36
generalized forces, 107, 153
generalized mass, 80
generalized stiffness, 80
Goland wing example, 144

I

inertial-reaction force, 82
influence functions, 12

K

kinetic energy
beam
assumed-mode lumped parameter, 106
definition, 83
bending and twisting, 88
finite wing, 140, 148
representative section, 114
torsion rod, 84

L

Lagrange's equation, 85, 89, 107
Langley Aerodrome, 4
Leibnitz's rule, 31
limit-cycle flutter, 164
linear elasticity, 5
Lockheed P-38 fighter, 180
lumped-parameter model
assumed-mode formulation, 106
flexibility formulation
bending, 92
coupled bending and twisting, 165
torsion, 93
stiffness formulation, 95
free-free beam with centre body, 104
mass matrix, 101

M

Maxwell's reciprocal theorem, 7

N

NASA X-29, 47
normal modes, 107

O

ornithopter wing, 24
orthogonality of modes, 80, 81

Q

quasi-normal modes, 154

R

Rayleigh-Ritz method, 36
reduced frequency, 117
representative section
divergence
definition, 28
example, 32
flutter
control surface, 174
definition, 113, 120
example, 147
rigid-body mode, 77, 78
rod-torsion elastic equation, 17
roll rate, 58
acceleration, 62
approximate equation, 62
steady, 58

S

semi-rigid wing model, 38
spoilers, 57
stall flutter
pitching, 163
plunging, 160
stiffness influence coefficient, 5
stiffness influence-coefficient matrix
definition, 7
segmented beam, 95
single uniform segment, 97
strain energy
aileron, 150
beam
assumed-mode lumped parameter, 106
definition, 14, 83
bending and twisting, 88
definition, 6
finite wing
bending, 141, 149
torsion, 141, 150
generalized, 8
integral, 13
torsion rod, 84

assumed mode, 36
definition, 18
strip theory, 31, 143, 152
Strouhal number, 160
structural damping
finite wing, 142, 152
parameter, 112, 114
representative section, 114
symmetrical modes, 78

T

Theodorsen function, 119
Timoshenko beam model, 97
torsional vibration, 67
assumed-mode solution, 84
cantilever rod, 68
free-free rod, 69
kinetic energy, 88

V

vibration frequencies, 69, 70, 74, 76
vibration modes, 69, 70, 74, 77
virtual work, 36, 37

W

weighing matrix, 34

FSC
www.fsc.org
MIX
Papier | Fördert gute Waldnutzung
FSC® C083411

Kolibri 360 GmbH
Ferdinand-Jühlke-Straße 7
99095 Erfurt, Deutschland
produktsicherheit@kolibri360.de